Technology
and Culture

Technology
and Culture

Allen W. Batteau
Wayne State University

WAVELAND
PRESS, INC.
Long Grove, Illinois

For information about this book, contact:
Waveland Press, Inc.
4180 IL Route 83, Suite 101
Long Grove, IL 60047-9580
(847) 634-0081
info@waveland.com
www.waveland.com

Photo credits
17 China © Oxford Science Archive / Heritage-Images / The Image Works; **30** Courtesy of U.S. Department of the Interior, National Park Service, Thomas Edison National Historical Park; **61 (top)** AP Photo; **61 (bottom)** AP-Photo/nap; **64** AP Photo; **76** Courtesy of Airbus S.A.S.; **83** AP Photo/Daniel Miller; **96** Courtesy of IBM; **113** Cevan Castle; **131** © Bettmann/CORBIS

To my wife, Susan Miller

Contents

Acknowledgments

Writing can be a marathon, a cross-country hike, or an endless journey. Writing this modest volume, by contrast, has been a sprint, an opportunity to bring together some thoughts that I have been working on for the past ten years. Along the way many students, friends, and colleagues have contributed to the ideas presented here.

Working with industry and government groups has been a useful challenge to keep theoretical concepts grounded in practical applications. Special mention should be made to Wizdom Systems, Inc., the Automotive Industry Action Group, the Air Force Materials Lab, the National Aeronautic and Space Administration, and many others for project opportunities in which anthropology could be engaged in solving technological problems. Appreciation also goes to the Aviation Safety Council of Taiwan, China, for the opportunity to apply concepts developed by my research team in a problem-solving context. Multiple grants from the National Science Foundation supporting my research are also gratefully acknowledged, as well as those from the Consejo Nacional Ciencia y Tecnología of Mexico.

Numerous international colleagues have provided opportunities to develop some of the ideas here. Among these are Alain Gras and Caroline Moricot at Centre d'Etude des Techniques des Connaissances et des Pratiques, Carmen Bueno and Marisol Perez at Universidad Iberoamericana, Maria Josefa Santos at Universidad Nacional Autónoma de México, Jing Hung-Sying at National Cheng Kung University in Taiwan, ROC, and Galina Suslova at the National Aviation University in Kiev, Ukraine. These, plus other participants in the workshop on Technological Peripheries (Mehdi Alaoui, Ricardo Dominguez, John Forje, Barbara Kanki, Mykola Kulyk, Saad Laraqui, Ben Mejabi, Ashleigh Merritt, Servando Ortoll, David Robertson, Thomas Wang, Lisa Whittaker). Closer to home, Marietta Baba, Dale Brandenburg, Donald Frey, Fred Gamst, Julia Gluesing, Brian Kritzman, Constance Perin, Charles Perrow, Bryan Pfaffenberger, Dennis Schleicher, Matt Seeger, and Stevan Weaver, have been valuable sources of stimuli and challenge. Special mention should go to Ann Jordan's assistance in the path toward publication.

My students as well have been a major source of ideas and sharp questions. Special mention is made of those past and present doing dissertations in technology and culture, including Dawn Batts, Kirk Cornell, Ricardo Garza-Miranda, Diana Gellci, Richard Krakowski, Elizabeth Nanas, Marlo Rencher, Laura Rodwan, Amy Goldmacher, Tara Eaton, Chris Miller, Brad Trainor, and Laurie Novak. A special mention must be given to my research team on technology and safety, Carolyn Psenka, Alex Perez, and Margaret Karadjoff, all of whom are undertaking pathbreaking work on the cultural contexts of technology.

Preparing this manuscript would have been far more difficult without the timely feedback of Lamees Sweis, whose insightful and informed comments foreshadow a promising career as an anthropologist of globalization. Final editing and proofchecking by Jeni Ogilvie at Waveland Press, and Samra Nasser, have done much to improve the manuscript. What few merits this book may have owes much to these; its more numerous defects remain my responsibility.

Finally, the support and companionship of my wife, Susan Miller, along this way, is but one of many reasons to dedicate this book to her.

About the Author

Allen W. Batteau is an auto mechanic, computer programmer, ethnographer, pilot, and university administrator. He is an Associate Professor of Anthropology at Wayne State University in Detroit, and Director of the university's Institute for Information Technology and Culture.

Prologue

Technology, and *culture,* for most of the public, evoke dramatically different worlds of meaning. Technology has to do with tools, highly engineered devices, computers, "high tech" industries, and a contrast with the industrial landscape of smoky factories and proletarian labor. Technology is clean, powerful, exciting, and a magical key to prosperity. In this view, there are *no* problems of contemporary humanity that cannot be solved with technology: diseased bodies can be made healthy, inefficient businesses and public agencies can be streamlined, borders can be secured with a "virtual fence," and improved communication will promote world peace. As naïve and starry-eyed as these statements may seem, they are seriously believed by many government and corporate leaders, even today.

"Culture," on the other hand, lives in a separate universe of discourse. For many, culture can refer to the fine arts—poetry, painting, drama—and advanced learning, "the best which has been thought and said," in the Victorian formulation of Matthew Arnold (1994[1869]:5). Culture, in Arnold's view, was the hard-won achievement of a lifetime of cultivation, reserved for the elite.

In contrast to this is anthropologists' definition of culture: a learned system of shared understandings that provides cohesion and a distinctive identity to a community. Culture, in anthropologists' view, is not an elite achievement but a democratic birthright shared by all. Communities and societies are distinguished by their unique cultures, even if the boundaries of any given culture may be frequently negotiated.

Over the past 20 years, as studies of technology regained respectability among anthropologists (Pfaffenberger 1992a), there have been three separate discourses of the relationship of technology and culture. The *utilitarian discourse,* associated with the historical materialism of

1

Marvin Harris, understands culture as the patterns or familiar grooves of human behavior, established in adaptation to the natural environment, given the toolkit at hand. Thus, Paleolithic hunters and gatherers, using stone tools, create one type of adaptation, focused on small bands and personalistic forms of authority, whereas agricultural societies, having mastered plows and animal traction, have evolved different sorts of institutions, including class differentiation, cities, and ruling classes. In this view, tools are understood primarily for their utilitarian qualities, for the manner in which they allow human groups to harness external sources of energy and bend Nature and fellow humans to their will. In this view, technology is the prime mover of history, and technological innovation, like Prometheus' fire from the gods, is the driver of cultural evolution. Although students of science and technology have largely abandoned this view of technological determinism, it is still found in several other academic disciplines, including some corners of anthropology.

A second discourse might be labeled the *materiality discourse*, which sees tools and artifacts in general as embodiments of cultural meanings, embodiments that members of a society use to order their society. The social meaning of a tool is as much a result of its production as of its employment (Dobres 2001) and is not distinctive from other material productions. Culture is a system of meaning in action that explains to us what tools, the environment, and fellow humans are, and what we should do about it. Culture is the meaningful constitution of practical interest that determines material logics (Sahlins 1976:207). Culture presents itself not only in myths and rituals but also in the material objects that, by encapsulating meaning, order social relationships. In this discourse, tools are but one more class of cultural artifacts, alongside and no more privileged than household goods or stylistic gestures in clothing.

A starting point for understanding the relationship of technology and culture is to appreciate that "tool" and "technology" are quite separate concepts. "Tools"—plows, digging sticks, knives—are about usefulness, and in most cases do encapsulate social relationships. "Technology" is also useful, most of the time, but condenses some new dimensions of cultural meaning: words, standards, connections, capability, insight, and aesthetics. Technology is both a large-scale system of imperative order and a collective representation specific to modern societies. Technology is a project of civilization, or more precisely civil-ization: bringing civil or corporate order to chaos. As Leo Marx (1997) observes, technology is a "hazardous concept" that contains within it many traps, in part because the term embraces a specialist's sense (referring to classes of devices) and a popular sense (referring to attitudes toward and expectations of awesome power and mystery [Gell 1988; Noble 1997; Nye 1994]).

At the risk of pedantry, therefore, I should state that when I use the word "technology" I am not using it in the unrestricted sense that most anthropologists use it. Anthropologists tend to use "technology" to refer to any toolkit, including those of societies lacking a written language. Determining what is primarily a "tool," as contrasted to some other sort of artifact, tends to make use of some of the presuppositions employed by European cultures regarding utility that I discuss in chapter 1.[1] *A fortiori*, bounding the concept "technology" (a necessary step if one wishes to make exact statements about "technology") is a hopeless task. The boundaries of "technology" are less semantic (implying clear distinctions) and more poetic (creating moods and expectations; [Burke 1938]). Thus, when I wish to refer specifically to objects of instrumentality, I will use the terms "device," "artifact," "tool," or perhaps "system," depending on which aspect (ingenuity, materiality, usefulness, or integration) I wish to emphasize. When I use the term "technology," it will be with a self-conscious intention of including the moods and motivations that are contained in the popular understanding of "technology."

In a third discourse, which I shall name the *managerial discourse*, "culture" is sometimes seen as the irrational, nonmanageable elements of organizational life. Any activity that is fully rationalized—an industrial assembly line, corporate management, laboratory research—is not cultural but rational, following the dictates of efficiency and the Laws of Nature. From this viewpoint, irrationality, for example workers' shirking on the assembly line or "casual Fridays" in many offices, is about culture; the assembly line itself, with its Tayloristic organization and mind-numbing pace of production, is simply the most efficient arrangement imaginable. When corporate managers take an interest in "culture" it is frequently out of a desire to fine-tune their workers' habits of the heart to promote further efficiencies.

Many people take for granted the class relationships and the vision of humanity that support this "irrational" view of culture. After all, it is established more firmly in managerial ideology and its literature (Deal and Kennedy 1982; Hofstede 1997; Peters and Waterman 1982) than it is in any empirical research into the cultures of the world. The predominance of managerial rationality in today's society, and the privileged status it affords to technology, creates several paradoxes, in our (cultural) understanding of technology, such as:

- Does technology, on balance, make a positive contribution to human happiness, freedom, and security? Or can this balance be maintained only by overlooking the wars, mass impoverishment, nuclear peril, and environmental devastation of assuredly the most technological of all centuries thus far, the 20th?

- Is modern technology qualitatively different from the technologies of earlier eras? If so, then is it possible to make general, nonethnocentric statements regarding technology?

- What is the substance of technology? Is it material? Words? Relationships? What sort of material, words, or relationships? Without careful specification here, "technology" can become a term like "culture" once was, a concept so all-encompassing as to be meaningless.

These are questions that will guide us throughout this book. Some other questions of technology and culture, such as the "productivity paradox" discussed in chapter 4, the role of the "human factor" in technology discussed in chapter 5, or the "two cultures" debate discussed in chapter 6, make it clear that some of the most urgent questions of technology-in-use are cultural, just as some of the most urgent questions of culture-in-practice are technological. My broad canvas should be suitable for a modern context in which "technology" has become as world-constitutive as "culture" once was.

Specifically anthropological questions, such as whether technology is an independent variable (a *primum mobile*, as Appadurai's language in *Modernity at Large* [1996] seems to suggest) or a social product, will help us gain insight into these issues. The view that will be developed in this book is that *modern* technology is so completely interwoven into the texture of *modern* society that the mutual evolution of modern technology and modern cultural forms cannot be usefully unraveled. Karl Marx stated that Watt's steam engine gave us the factory; historical precedence would suggest that the factory gave us the steam engine.

Questions such as these possess specific urgency because of pride in the place given to *innovation*, the continuous and continual effort at self-transformation in our contemporary society. Innovation is the socialization of invention, turning novelty into social value. In 1903 the airplane was an the invention of the Wright brothers.[2] For the next 11 years the airplane was a novelty, acquiring some limited utility in the World War I as a weapons and observation platform. (Whether military weapons have social value is a question that will linger with us until chapter 6.) In the 1920s, airplanes were primarily used for entertainment; their other uses were dependent on government subsidy. Only in the 1930s did aircraft manufactures find the right combination of speed, capacity, reliability, route topology, and customer demand to make commercial flight a viable *innovation*.

Invention can be measured by patents issued, which have increased at a steadily accelerating rate through the entire 20th century. Innovation is more difficult to measure, inasmuch as there is little consensus on what factually constitutes an innovation; a new coat of paint, a new style of tailfin, or a new style of tailoring might all legitimately be innovations.

The economist Joseph Schumpeter (1942) characterized the process of innovation as the "creative destruction of capital." Following are some examples that illustrate this concept. After America's Civil War, railways were the backbone of the American economy. They were the corporate giants that bestrode an entire continent, and their freight and passenger business dominated the nation for more than a half-century. Then their freight business was driven into receivership by trucking on interstate highways and the passenger business by airlines and automobiles. As another example, newspapers were leading businesses in every American city in the 18th, 19th, and 20th centuries, and their publishers were pillars of the communities they served. At the end of the 20th century the core of newspapers' business—news and advertising—was undercut by television (especially cable) and the Internet. One final example is the *Encyclopedia Britannica*, an authoritative resource for ages, which sat proudly in the reference section of every library and on the bookshelves of many aspiring families, until it was "blown to bits" by an inferior product, *Encarta*, encyclopedia software given away for free by an upstart company named Microsoft (Evans and Wurster 2000).

If culture is about permanence, stability, and cultivation—the best which has been thought and said—then technology is about innovation, upheaval, and creative destruction, the most awesomely efficient which has been made and done. In a globalized world this contest between technology and culture assumes Titanic proportions. Understanding how culture creates, fosters, inhibits, and reframes technology, and how technology supports, undermines, and ultimately redefines culture, is the objective of this book.

Notes

[1] Anthropologists often equate their "scientific" understanding of a concept with an "etic" definition, in contrast to a popular or "folk" understanding, which is seen as "emic" (as in the contrast between phonemics and phonetics). Thus, the popular "gee-whiz" understanding of technology is seen as an emic concept, whereas our possession of "true insight" is etic. This misplaced privileging of specialists' taxonomies is discussed (albeit briefly) in chapter 6. It is sufficient to note here that both the specialists' and the popular understandings are cultural constructions.

[2] More accurately, the Wright brothers invented airborne control of longitudinal motion, or roll. Other inventors had also been experimenting with heavier-than-air flight for decades and quickly seized on the Wrights' invention of wing warping. The airplane, properly speaking, was the result of a collective effort in Great Britain, France, Germany, Italy, the United States, and Brazil, among hundreds of experimenters.

Chapter One

Tools, Technologies, and Cultural Evolution

Before there was thinking, there was making. Before *Homo sapiens,* "thinking man," evolved the enlarged cranium we associate with thought, there was *Homo erectus,* formerly called *Homo habilis,* "capable man," the first member of the genus *Homo,* walking erect, grasping tools, and using fire to warm dwellings and sharpen weapons.

What does it mean to be human? The *meaning* of humanity, a question at the core of the anthropological project, has animated poets and philosophers at least since Sophocles' wide-eyed wonder at man's ability to cross the ocean's waves and wear away the earth.[1] In the Age of Faith (the Middle Ages), both humanity and civilization receded in significance, as church fathers directed learned attention toward life after death. In the Renaissance, with classical learning reborn, humanism replaced scholasticism, and Man once more assumed the center of inquiry.

As 15th- and 16th-century travelers began to discover new continents, and discover new, strange, two-legged beasts, the question of the meaning of humanity shifted from philosophy to a matter of state. Did these beasts have souls? Could they speak a language? How might they be governed? Shakespeare's shipwrecked mariners in *The Tempest,* led by Alonso the king of Naples, replayed this discovery, wondering whether Caliban, half-human, half-beast, was truly human. Anthropology's hard-won conclusion, that humanity in our unity is defined by shared capabilities in sociality, language, tool using, and sign making, constitutes an advance in culture no less momentous than Watt's steam engine constituted an advance in industry. *Technology* adds to these advances historically situated assemblages of *technê* (craft, skill, or art)

7

and *logos* (representations, state power, and language), embracing arrays of tools, documents, social connections, and images.

In this chapter, we will examine the relationship between tool-making, technology, and cultural evolution. Rather than attempting to reprise hundreds of years of archaeological research, thousands of years of history, or millennia of human evolution, I will organize our thoughts around four questions that animate this book and that the archaeological record lends some privileged insight into. These questions, in some respects, have already been answered, either in the popular imagination or in the conclusions of professional archaeologists. Yet, I will attempt to enlarge both of these in order to achieve a more consistent understanding of the relationship of technology to culture.

First, in what manner does the *character* of earlier tools and technologies differ from those modern tools and technologies that we are familiar with? Although this book focuses primarily on modern technology, I wish to make clear in what way modern technology is a subset of a larger class, technology, yet adds some unique characteristics not found in earlier toolkits. Most notably, the greater *social* complexity of modern technology amounts to a difference in *kind*, suggesting a new dynamic in the evolution of sociotechnical systems.

A second question is the relationship between technology and language. Although a substantial body of research (and speculation) has gone into the use of language and tools among early hominids, I will confine myself to the conclusion that technology is half *technê* and half *logos*, and that this marriage of words and tools creates a zone of autonomy for the world of tools (Gibson 1991).

The third question is the relationship between technology and material culture. The commonsensical view sees technology as characteristic of devices—iPods, internal combustion engines, jet airplanes. Yet before the 20th century, most of the works on technology, from Vitruvius (a Roman architect and engineer, born c. 80 BC) to Bigelow (a medical doctor born in 1787), primarily described roads, walls, bridges, dwellings—things that *contained or connected* members of society. Actor network theory, a view of technology that emphasizes associations, places less emphasis on the artifact *per se*, and more emphasis on the networks of artifacts, social groups, and other formations within which technological artifacts are embedded. How material culture can thus create or reinforce social boundaries becomes an important technological question (Miller 1987).

Finally, ambitiously and inadequately, I wish to open the question of whether technology overall has made a positive contribution to human happiness, freedom, and security. The unreflective answer would be "yes, of course it has." At the close of a 20th century of world wars, holocausts, and nuclear devastation and at the beginning of a 21st century of terrorism, ethnic cleansing, and environmental hazards, one must pause before accepting this unreflective conclusion.

FROM SIMPLE TO COMPLEX, AND BACK AGAIN

Village- and band-level societies are defined in part by their lack of differentiation. Within their environments, they typically display high levels of ingenuity in adapting their resources to their cultural objectives, and high levels of sophistication in their myths and rituals; this ingenuity and sophistication suggest that these societies should not be called "primitive." An earlier view, that these societies represent intermediate stages along humanity's advance from cave dwellers to city dwellers must likewise be abandoned in favor of regional or world-view perspectives that stress the interrelationships between highly stratified and egalitarian communities. So-called "primitive" societies more plausibly co-evolved as part of larger regional complexes containing both egalitarian and stratified sectors.

Regardless of what conclusions one draws regarding such communities' cultural "level," it is inescapable that their toolkits are less complex, their social range more constricted, and their social orders less differentiated than those of modern cities. It is often thought that technological "advance" creates social differentiation: just as the steam engine gives us factories with a division between owners and workers, so, too the automobile creates a division between cities and suburbs and opens up endless possibilities for residential selection. With equal plausibility, though, one could argue that social differentiation engenders technological complexity: there is nothing in the laws of physics that requires that in the United States three automotive manufacturers produce 10 nameplates with 97 models, thousands of combinations of body styles, and millions of possibilities of option content. Nor is it in response to any electromechanical necessity that certain of these combinations become objects of cult-like attraction. Automobiles such as the Chevrolet Corvette, the Ford Mustang, and the Dodge Viper all have local clubs, rallies, magazines, and other artifacts of loyal devotion. As such, they qualify as *technototems*, in David Hess' phrase (1995:54ff.), technological embodiments of social distinctions.

This variety is a consequence of *technological exuberance*, the tendency of all inventions and inventors to spawn variations on the basic theme. In the 19th and 20th centuries, airplane designers experimented with thousands of configurations of wings, propulsion, and control configurations, before settling on a few basic designs. Some of the early airplanes had five wings—one even had 20. From a 21st-century perspective, many of these designs are quite amusing, but at the time, they were serious enough for the experimenters to hazard life and limb and substantial fortunes in the effort. Designers of bicycles and automobiles likewise experimented with a lush profusion of structures, propulsion mechanisms, and control concepts. Eventually the exuberance settles down to a few basic themes that the environment can support. The environment, of course, is not only the physical environment but also the social and the epistemological environment. These themes become *attractors*, templates for subsequent innovators.

Attractors such as hot cars co-evolve out of a niche-seeking behavior of technological exuberance, an identity-seeking behavior of social distinction, and a permanence-seeking behavior of cultural images. An occasional coalescence around a specific combination of object, image, and identity may seem strange to outsiders, but to those on the inside— the members of a motorcycle club, for example—*it's who we are*. So much identity is invested in the object and the image that one dare not question it. Nor is this reserved for transportation device culture: nearly everyone in contemporary society has objects into which they pour their identity: family heirlooms, for example, or kitchen appliances, or suburban faux mansions. What differentiates modern society from earlier societies is the degree to which these identity objects use technological means to package a broad array of social values—a four-wheeled Mustang on today's highways, in contrast to a four-legged mustang on the Great Plains a century ago.

Identities and identity objects simplify social life. Like a sonnet, or a painting, they compress a broad array of experience into a few appealing images, organized in a conventional form and easily grasped by the reader or the rider. Like good industrial design, identities package complexity within simplicity, turning *useful* objects into *usable* objects, and usable objects into *meaningful objects,* the experience of which has meaning well beyond just getting the job done. The iPod, for example, is about more than just putting a lot of music into a small container; small containers, after all, come in many shapes, textures, and colors. With its slim design, soft colors and rounded edges, the iPod has emerged as a 21st-century embodiment of youthful engagement.

The meanings invested in objects extend well beyond *utility*, the orthodox economists' measurement of the desirability of some good or service. Utility reduces all value to a single scale and attempts to assign an interval or ordinal measurement to determine the rationality of peoples' choices in the goods and services that they buy and sell. While this works well for bulk commodities, such as pork bellies and crude oil, it is demonstrably nonsensical for those objects and expressions to which a culture assigns its deepest, most profound value. What, after all, is the utility of a sonnet? Or a lover's sigh? Or a scrap of cloth rendered in red, white, and blue? Many have died for these, behavior that from a standpoint of economics would be dismissed as "irrational." Understanding that technology's value extends well beyond utility helps explain some of its seeming irrationalities, including the expectation that every problem has a technological solution.

This distinction between family heirlooms and technototems is less one of origin and more one of the meaningful context of the object, between kinship networks and sociotechnical networks. To say that an object is a family heirloom enmeshes it in a web of kinship, life history, and landscapes of memory. To say that an object is technological connects it to a separate universe of meaning, of equal cultural interest: magical capability, warping of human scale, and heightened expectations. Technological objects are uniquely integrated with other technological objects: To safely fly an airplane requires radios, navigation aids, airports, air traffic control, flight rules, a system of fuel supply, and a superstructure of government certification and supervision. Thus, when we speak of "technologies" it is more accurate to speak in terms of *sociotechnical networks*. Every useful technology is embedded within a sociotechnical network and can neither function nor be understood nor even exist apart from that network.

The consequences of this complexity are that modern sociotechnical networks are more brittle—that is, less robust, less adaptable to changing circumstances—than simpler systems. This was conclusively demonstrated on September 11, 2001, when 14 terrorists with box cutters overcame the defenses of three flight crews, two airlines, the FAA, and the U.S. Air Force. By contrast, a handful of passengers on American Airlines Flight 77, armed only with cell phones and courage, defeated a fourth band of terrorists, succeeding where squadrons of scrambled F-15s and F-16s and the entire NORAD air defense network failed (National Commission on Terrorist Attacks 2004; Scarry 2003). Less dramatically and more pervasively, it is demonstrated every day when we adjust our lives to increasingly demanding technology.

Social complexity goes along with technological complexity. This is reflected in functional specialization in today's economy: the Census Bureau's *Dictionary of Occupational Titles* lists 12,740 distinct occupations and nearly 10,000 Standard Industrial Classifications. Social complexity is also reflected in today's diverse memberships, styles of life, and cultural media, all of which technological development enables and supports. Social differences, whether of race, class, or nationality, in Peter Fritzsche's phrase, are given shape and sturdiness by technology (Fritzsche 1992:3). Social complexity also raises questions of a community's governability, an issue that will be considered more closely in chapter 7.

Complex, tightly coupled systems, by definition, are less adaptable to changing environmental circumstances than are simple systems. The Ford Motor Company, for example, produced its first car, the Model A, in October 1903, four months after going into business. By contrast, car companies today require from three to five years to produce a new car, even as consumer tastes churn and skip and dance around yesterday's styles. The Ryan Aircraft Company needed only three months to pro-

duce and prove the NYP-211, The Spirit of St. Louis, from the first drawings in a shed in San Diego to the May 20, 1927, departure for Paris. By
contrast, the Boeing 787 Dreamliner was more than 10 years in development. Similar stories can be told for every advanced technology.

As technological systems become more complex, and as long as we
invest them with expectations of awesome performance and inevitabil-
ity, they will exert a growing dominance over everyday life. The costs,
performance deficits, and restrictions on freedom imposed by complex
technologies make it clear that the evolutionary path from simple to
complex is not nearly as straightforward as is that of individuals *choos-
ing* to burden their lives and restrict their freedom. More accurately,
the evolutionary trajectory should be characterized as a self-reinforcing dynamic where asymmetries of power and connectedness, including
those of gender, ethnicity, and class, acquire shape and sturdiness by
being invested in technological systems. These asymmetries return to
our daily lives as technological imperatives having great force by virtue
of having the Laws of Nature and the attachments of identity invested
in them. These attachments of identity—technototemism—are what
enable nontechnologists to navigate through a world that is increasingly a State of Technology.

FROM SERVANT TO MASTER

A fundamental debate in studies of technology and culture is the
degree to which technology is an autonomous force, resistant to social
influences, making demands to which society must adapt. Karl Marx,
V. Gordon Childe, and Leslie White (1949) saw toolmaking as the prime
mover in history, a progressive accumulation of instrumental knowledge giving tool-equipped humankind greater mastery of the natural
environment. Although most scholars have abandoned this position of
technological determinism, in policy circles and the popular imagination, technological imperatives must be respected: if it *can* be done or
built, it *must* be. In wartime, technological arms races make some
sense, albeit from a one-sided perspective. In the absence of open hostilities, such justifications are far weaker. Space exploration is no
longer justified in terms of national defense; instead, it is identified
with the "endless frontier" of the scientific pursuit of knowledge. Thus,
technologies (or sociotechnical complexes such as the space program),
unlike tools (such as hoes and horse-drawn plows) *are* autonomous and
possess trajectory, momentum, and agency. One thinks of *tools* as existing to serve humanity, and not the other way around.

Tools began to free themselves from local circumstances and
evolve into technologies when their makers or maintainers began cre-

ating autonomous representations of the tools. An autonomous representation is a picture, a song, a dance, a legend, or a written document
that describes the tool. Cave paintings at Lascaux, in southwestern
France, show the spearing of bulls, perhaps one of the earliest known
representations of tool using. Certain Inuit bands have dances in which
they re-create the harpooning of seal, both as celebration for the hunters and instruction for the young. Homer's *Iliad* lists the ships that
sailed into the waters surrounding Troy, more in terms of the kingdoms
that dispatched them (a social context) with only passing reference to
their physical characteristics ("long black ships"). Chanted odes, carv-
ings on the blade of a knife, or dances around a whale-oil lamp make it
clear that these tools are social objects; none of these representations,
however, has the portability, flexibility, and ease of use of a universal
component of every technology that deserves the name—that is, a written description. In the absence of a written description and drawings,
objects of ingenuity are just clever devices, meaningful for their makers, useful in their village, but denied the ticket of full admission into
social circulation. Bruno Latour (2005) and actor network theory make
clear that social circulation is a defining part of technology, yet in their
focus on people, artifacts, and relationships, they devote only passing
attention to the importance of the standards that extend it.

Vitruvius, in *De architectura* (*The Ten Books on Architecture*, 2004),
supplied what is probably the earliest known Western instruction in
engineering practice. He described methods and principles for building
houses, temples, walls, public buildings, and machinery. He based his
principles on timeless facts, such as the symmetry of the human body,
instead of on personal preferences. Therefore, *De architectura* deserves
to be seen as one of the earliest documents of technological standards. In
China, a division between the learned (those who wrote about
monumental construction), and the illiterate artisans (those who
actually did the work) seems to have prevailed, leaving little instruction
written by the actual creators. Nevertheless, from the Qin period (221–
206 BCE) forward there are documents describing large-scale works such
as walls, roads, and canals (Needham 1954). Ever since then, tools and
instruments that have evolved into classes of devices have followed or
acknowledged previous standards (even if only to explore their limits),
and have been accompanied by documentation that was no less critical
than the artifact itself. For example, years before their first successful
flight, the Wright brothers consulted the notebooks of the German
experimentalist Otto Lilienthal (1893), Langley's *Experiments in
Aerodynamics* (1891), and the *Aeronautical Annual*, published from
1895 through 1897. Like good engineers, they made careful drawings of
their flyer, and kept careful records of their results, so that others did not
need to journey to Kitty Hawk or Huffman Field outside of Dayton to
learn from them. After the results of the Wright brothers' demonstration

at Hanaudières, outside Paris, were published in *Le Figaro* in August 1908 their aeroplane flew around the world. Standards guide technological innovation, and documentation perpetuates it.

Standards and documentation also create for technology a zone of autonomy, freeing it from the tyranny of the here and now. If all that an engineer knew about carburetors came from handling the device and observing it mixing gasoline and air on top of an internal combustion engine, then experimenting with new configurations would be quite difficult and time-consuming. The engineer would need to remake the device every time he or she wished to try any variation on the original design. However, having drawings, specifications, test documents from other configurations, plus tables of vaporization and fuel flows, our engineer can successfully *imagine* a new carburetor before actually *building* it. Thus, documents showing drawings of details about the carburetor, like every other technological object, allow it to travel much farther and faster than an actual physical carburetor ever could.

We might consider that *De architectura* marks the beginnings of technology, although the world would have to wait nearly 2000 years for it to be named as such. For most of history, books such as *De architectura* or *De aquaeductu urbis Romae* by Frontinus were considered manuals of the useful arts, of great importance to engineers and tradesmen, but not part of a society's scientific or learned heritage. Occupying a lower rung on the society's hierarchy of knowledge, their development could more easily be guided and controlled by politicians, priests, and patriarchs. The dynamic growth of technology that we have witnessed in the 20th century would have to wait until technology came into its own right as a respected branch of learning.

FROM ISOLATION TO CONNECTION

For an object to be considered "technology," as contrasted to a Rube Goldberg contraption or some object of local ingenuity, it must enter into social circulation. Otherwise, practically anything that anyone devises—a pile of stones to mark a trail, or an improvised patch on a leaky roof—must be considered technology, and the term loses all analytic value. "Technology" has additional characteristics as well, but instrumentality and social circulation are at its core. Also at the core is the fact that technologies embody some level of engineering, meaning a knowledgeable arrangement of forces and materials. The difference between an improvised barrier that lasts a season and an engineered wall that lasts a thousand years is evident for anyone to see: less visible, except in the result, are the knowledge and craftsmanship that went into the construction of the latter.

If we consider instrumentality, social circulation, and engineered effort as definitive of technology, then we can survey the artifacts of the ancient world and decide which of these represent technological achievements. The walls of Jericho might be the earliest of these, although the fact that they came tumbling down suggests that in 3500 BCE, the art of wall construction had not advanced very far. A thousand years later the great pyramid at Cheops displayed a technological achievement that after 4,000 years is still standing. On the other hand, plows and other agricultural implements probably should not be considered technological: they tend to be adapted to local situations, and I have been unable to find any written document describing standards for plows, in contrast to *De architectura*, which does describe standards for walls and dwellings. The relative social status of agricultural laborers and city dwellers may supply sufficient explanation for this greater attention paid to the artifacts of the latter than the former. On the other hand, a fresco from a tomb at Thebes, c. 1420 BCE, showing land-reclamation techniques, might be one of the earliest representations of agricultural tools or instruction in their use.

When one surveys premodern technology, both in the ancient world and in the Middle Ages, productivity and domesticity do not have a commanding role. Other objectives, such as warfare, monumental displays, and social order, figure more prominently. The first engineers were actually military engineers, building fortifications, tunnels, ballista for hurling objects, and warships. After warfare, monumental displays—whether the pyramids and Sphinx of Egypt or the ziggurats of Mesopotamia—are prominent. Later, temples and palaces, such as Karnak in Egypt, served more for imperial display than for worship or humble residence.

The taken-for-granted association of technology with power has created a view of technology that associates it with masculinity and heroic achievement. Early in the Mercury space program, candidate female astronauts exceeded their male counterparts in all respects of mental, psychological, and physiological qualification, yet they were not chosen to go into space: they had "the right stuff, but the wrong sex." Before and since, technology has been a "guy thing," distorting both the understanding of technology and its contribution to reinforcing gender distinctions. Consequently, those activities that are predominantly male are featured in technological accounts, whereas those that are predominantly female are overlooked. Like class, status, and rank, gender differences are inscribed in technologies in ways that reinforce and obscure their social construction. (For more gendered views of technology, see Cowan 1983; Wajcman 2000, 1991.)

Military and monumental technology have one feature that limits their dynamic potential: they are unproductive. They require constant infusions of resources, whether taxes or forced labor, infusions that have caused the collapse of many empires. Production technology, by contrast, whether in agriculture or manufacturing, can create social surpluses enabling a self-reinforcing dynamic of growth and evolution.

Other ancient technologies that did create growth potential include containers, such as dwellings and city walls, and connectors, such as roads and aqueducts. Just as good walls make good neighbors, they also make good citizens, both inside the walls of a dwelling and inside the walls of the city. The combinations of containers and connectors created of social autonomy within the walls and integration along lines of communication were directly related to the growth of the empire: better roads meant better communication with the provinces (including sending in the Roman army), and better communication meant better tax collection for the benefit of Rome. Roads further acquire value by being part of a network of roads: through a dynamic of network externalities, each *mille passum* (literally, "thousand paces" but loosely translated as "mile") and every intersection added to the road network increases the value of the entire network. For those enmeshed within the network, everyone benefits from this dynamic of self-replication.

Instruments of communication likewise received imperial attention and required standardization. Coins, for example, communicated the name and likeness of the emperor at a time when there were few other mass circulation media. Inca pottery was used by the Inca emperors to communicate across the realm (Bray 2000). In all of these cases, we find the tokens of standardization, engineering skill, and social circulation that I have used to define technology in a strict sense of the term.

What we do not find is an all-pervasive presence of technology in citizens' daily lives. Soldiers and tax collectors might work with technological objects every day, but for the mass of the population, "technology" occupied only a corner of their lives. Craftsmen worked with tools every day, producing objects both of great usefulness and beauty from customary designs, but this was within a workshop tradition, in which a master established his own designs, methods, and standards. Written documents were not used, because in all likelihood the craftsman was illiterate and quite possibly a slave. The authority of the master artisan or craftsman was a personal authority and did not extend beyond his workshop and his apprentices. The rarity of documents such as *De architectura* attests to the lack of a strong, written tradition in the useful arts. Only when the useful arts were joined to state purposes, whether tax collection, road building, or weapons of war, was it sufficiently important to create documentation. What we now call "technology" was a branch of the arts, alongside painting and poetry, maintained through an oral tradition.

In this world there were many ingenious inventions, particularly around vexing problems such as irrigation. The shaduf, for example,

Raising water by means of a shaduf, China 1825–1835. The shaduf is a simple device with a bucket attached by a rope to one end of a shaft and, in this version, a counterweight at the other. The shaft is pivoted on a pole. The shaduf is used to lift water from a well or stream.

shown in the illustration above, a counter-weighted lever for lifting water into an irrigation canal, is found all over the Near and Far East, even today; its simplicity and local adaptability make any thought of "standards" for shadufs beside the point. By contrast, Archimedes' screw (shown on the following page), also an ingenious device for the problem of raising water, is more complex, and as incorporated into numerous pumps all over the world, is the subject of written standards. We might consider the shaduf a prototechnology, and Archimedes' screw a fully developed technology with a theoretical base and copious written documentation. Many other examples of classical ingenuity, from chariots to ships to glassblowing to water mills, could equally be given, illustrating the separate trajectories of production and connection devices.

The master of classical technology and the grandfather of modern, Western technology was, of course, Leonardo da Vinci (Florence, 1452–1519), a painter and inventor whose numerous inventions anticipated

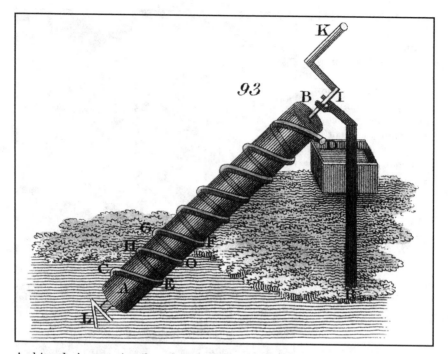

Archimedes' screw. Attributed to the Greek mathematician Archimedes in the third century BC, the Archimedes' screw consists of a screw inside a hollow pipe and was used to lift water from low-lying places to higher elevations for irrigation. It was also used to lift water from mines and ship bilges. It forms the basis for many modern pumps, and as such is the subject of several written standards.

many objects that are used even today. It is not known if his flying machine ever actually flew; Leonardo's statement, "For once you have tasted flight, you will walk the earth with your eyes turned skyward," suggests he did. The substantial body of drawings and documentation that da Vinci left behind have been both an inspiration and a practical guide to inventors and innovators ever since.

FROM THE STATE OF NATURE
TO THE STATE OF TECHNOLOGY

In 1651 the English philosopher Thomas Hobbes, one of the fathers of modern political science, wrote that in the absence of civil government, lives were "nasty, brutish, and short" (1668[1994]). Three centuries of empirical anthropology, motivated by actual rather than literary New World exploration, have conclusively demonstrated that

Hobbes' State of Nature was and is totally imaginary—a dystopia created to justify contemporary institutions. As Marshall Sahlins has described in "The Original Affluent Society," band- and village-level societies are frequently marked by ample leisure, nutritional adequacy, and rich family lives (Sahlins 1972). Jean-Jacques Rousseau, similarly, in his own State of Nature (which proclaimed "men are born free, but everywhere they are in chains") had an imaginary creation, only in his case for purposes of indicting rather than justifying contemporary institutions (Simpson 2006). If the State of Nature is totally imaginary, likewise is the State of Technology, that utopia or dystopia created by techno-enthusiasts and science fiction writers either to promote a gadget-filled Nirvana or to warn of its perils. Between these two imaginary extremes, can we make any statements regarding the progressive consequences of technological development for the human condition?

I have established that *technology* in a strict sense of the term embraces increasing complexity, increasing autonomy, and increasing connectedness. To the extent that these characteristics are cumulative and self-reinforcing, we are justified in understanding technological development as "progressive." Complex systems become more complex as energy is added to them (Holland 1995; Kauffman 1995); network-building (or system building, as we will see in the next chapter), as it creates greater value in networks, is likewise self-reinforcing.

This, however, makes no judgments as to whether such "progress" is actually an improvement of the human condition: the fall of the Roman Empire, for example, like that of many other empires, was progressive: it was not so much a cataclysmic event as it was a self-reinforcing deterioration spanning several hundred years. Some writers, such as Jacques Ellul in *La technique* (*The Technological Society*, 1964) and filmmakers, such as Fritz Lang in the 1927 movie *Metropolis*, have insisted that technological development also presents a process of self-reinforcing deterioration, an enslavement of all of humanity either to the machine (*Metropolis*) or to the soulless forces of managerial rationality.

It is only in the modern era that the idea of progress has achieved currency as something that is both good and inevitable, a cultural shift that has been debated by philosophers and historians for more than a century (Bury 1932). Earlier philosophies of history either stressed a lack of change (Buddhism), a cyclical view (the Hindus), or a declension from a Golden Age (Erasmus). Only since the 18th century have philosophers seriously argued that the direction of history was one of inevitable improvement. The association of this inevitable improvement with the development of ingenious devices became a national religion only after the French and American Revolutions.

Can any of this be justified? Is the world on a trajectory of increasing freedom, increasing abundance, and increasing security? Or are we headed in the direction of increasing coercion and insecurity? The relationship of technology to freedom and slavery, to abundance and scarcity, and to security and hazard, should be the proof standard for any technological development.

In the first regard, no honorable person would suggest that the disappearance of slavery is anything other than good. Although slavery has not disappeared, it is outlawed in every country around the world and survives only in lawless regimes such as Sudan, illegal sweatshops in Los Angeles, or brothels in Bangkok. Historically, we can make the distinction between slave societies (in which broad stretches of production and civil order were organized under conditions of legal unfreedom) and societies with slaves (where household slaves were present but economically marginal). This distinction suggests separate dynamics for the competition of slave labor with machinery. Slave labor can either be intensified with technical developments, or its disappearance can be accelerated (Turley 2000).

The dialectic between technology and servitude animates much of our story. Slavery as the fate of captives in war has existed ever since humans invented warfare, although the legal status, the racial marking, and the eventual fate of slaves in specific societies have varied enormously. Societies that disdained manual labor, whether the ancient Greeks or the Antebellum South, had little need for technological advancement: there were always slaves to do the tedious tasks. As long as labor is cheap and freedom is a privilege, technological advancement is unnecessary. Contrariwise, some of the greatest advances of industrial technology coincided with 20th-century restrictions on the exploitation of labor—whether the elimination of child labor or the 40-hour workweek. When free labor commands a minimum wage, and its hours are restricted and its injuries are compensated, then and only then does capital have the incentive to replace people with machines.

The contribution of technology to abundance is more problematic. There is no question that technological advances in farming and medicine have improved health and nutrition, although in a world of increasing obesity and escalating medical costs, it is questionable whether these improvements can or should be sustained. The world may have reached the point where further improvements in health and nutrition depend more on institutional—that is, cultural—reform than on technological advance, and that further technological developments might give us greater abundance of pollution and infectious diseases. Advances in transportation mean that infectious diseases travel further and faster, as the SARS epidemic in 2003 illustrated. Similarly, increased ease of mobility displaces population pressure away from villages where it was managed for millennia, and onto vast urban agglomerations, where mass poverty is the norm.

The contributions of technology to security are less ambiguous. On balance, there are none. The possibility of destroying a city with the push of a single button did not exist before 1945; the ensuing arms races have created a more dangerous world in the 60 years since. Technologies of warfare and control have given shape and sturdiness to 20th-century imperial ambitions, in ways that earlier empires could only dream of. Resistance to empire, whether in the form of piracy and banditry in earlier eras or terrorism in our own, likewise now acquires greater potential by being equipped with the entire arsenal of modern warfare technology, save nuclear weapons. This one exception may be only a matter of time (Allison 2004). Earthquakes and floods can be mitigated with advanced sensors and well-constructed levees, resulting in a perception of security that encourages people to build their homes in seismic zones and floodplains, allow public officials to neglect the levees, and assumes that protection from natural hazard is now a matter of "personal responsibility." Although technology can make single buildings or single gated communities more secure, this security cannot replace a sense of civic responsibility.

Perhaps the best one can conclude is that technological development creates opportunities for improvements in human happiness, freedom, convenience, and security, but with the same strength creates new opportunities to degrade the human condition. In a technological society, there is no longer the same justification for slavery as there was in ancient Greece or Rome, yet sweatshops are part of the modern industrial economy. In a technological society starvation should no longer exist, yet complex and tightly coupled arrangements of distribution, of agricultural policy, and of population policy, repeatedly fail. In an era of ubiquitous and instantaneous communications, international misunderstandings ought to be attenuated, yet many partisan ideologues leverage the technologies of communication to multiply misunderstandings. *Tools* are merely implements, innocent of the purposes to which they are put. When a tool enters social circulation as a technology, it picks up the values, social projects, and ultimate purposes of those who introduced them, giving those values and purposes a shape and sturdiness they would otherwise lack. How technology has formed and added force to the cultural values and projects of the modern world is a subject to which we now turn.

Note

[1] Rather than improve on Sophocles, "Wonders are many, but none is as wondrous as man"; this will be one of the few places where it is appropriate to maintain a gendered reference.

Chapter Two

The Invention
of Modern Technology

Human desire throughout most of history has exceeded human capabilities. For thousands of years, ever since those who *had* began living alongside those who *worked*, in the cities of ancient Sumer, Egypt, and Mesopotamia, dreams of unlimited wealth, leisure, and pleasure animated men and women. King Midas, the Sorcerer's Apprentice, and Aladdin's magical lamp are but three examples of fables where magical means could be used to achieve endless worldly wants. The Industrial Revolution, and its tools (steam engines, electricity, machinery) and institutions (factories, railways, distant communication), turned the dreams of millennia into the stuff of everyday life.

The technologies that we are familiar with today are products of the Industrial Revolution. The objects that we think of as technological are useful (sometimes), but just as importantly, they connect us to others. They connect us directly, as with telephones, televisions, the Internet, power grids, and transport devices, and indirectly in terms of their basis in a community's agreed-upon technological standards. The Industrial Revolution enlisted new sources of physical power for humanity, which through technology was transformed into social power.

Humans, of course, have been using tools for hundreds of thousands of years, and in the Middle Ages there were many learned philosophers, artful magicians, and skilled artisans. However, prior to the Industrial Revolution, no one joined natural philosophy, magic, and the useful arts together and characterized them as "technology." Rather philosophy, magic, and the useful arts proceeded on separate and unconnected tracks. The Greeks and Romans had impressive scientific and engineering achievements, but usefulness was considered to be beneath the dignity of gentlemen. In the Middle Ages, philosophers were clerics, men (mostly) of the Church and of great social esteem. Magicians practiced their black arts in the basements of castles, and artisans were men and women of great skill but very little formal learning, mostly illiterate. The fusion of *techne*—embodied craft skill, acquired over a lifetime of practice—with *logos* or knowledge, learned insight into the mysteries of the universe—would not appear in the West as a branch of learning before the 18th century (e.g., Bentham 2007[orig. 1827]; Bigelow 1829).

In this chapter, I describe the events and arrangements that gave rise to modern technology as a class of artifacts infused with awesome powers and expectations. I describe the role of system builders such as Thomas Edison, Alexander Graham Bell, and Walter Folger Brown in transforming numerous local manufactures and utilities into large-scale systems requiring large-scale agreements on standards and purposes. This transformation, which I will call the "technological turn," occurred at different points in different industries; it marks an epochal change in which technologists assume a commanding role and industrialists stitch island communities together into national systems.[1] Technologists create both these large-scale systems and the engineering standards on which the systems are based. Standards, whether frequency bands for radio or current voltages for electricity, express a collective will. This collective will distinguishes technologies from merely local tools. The world-transforming impulse behind this technological turn is apparent in the aesthetic most closely associated with technology, Modernism. Modernism, and its cousin modernization, captures the cultural and social transformations that liberated technology from parochial restraints and sent it around the world (Berman 1988; Harvey 1990). The joining of large-scale usefulness, collective will, and modernity created Technology as we understand it today.

MODERN TECHNOLOGY IN AMERICA AND EUROPE

What we now understand as modern technology was the product of a specific set of cultural and social circumstances centering on Europe and the United States in the 18th and 19th centuries. The Industrial Revolution gave rise to technology, the joining of *techne* with *logos*, as a discipline and later as a named class of artifacts. Most historians date the Industrial Revolution to James Watt's perfection of the steam engine (1765) over earlier efforts by Savery (1698) and Newcomen (1712). This development represents an early bringing together of craft skill and academic learning. A skilled tradesman, an instrument maker building and repairing scientific devices at Glasgow University, Watt had the occasion to make the acquaintances of professors there and discuss the operation of the devices he repaired. When a professor asked him to repair the Newcomen engine used in a natural philosophy class, Watt had the insight that a separate condensation cycle would make the engine far more efficient. Watt's engine, of course, would transform the world only when it was installed to create new devices and new forms of production: tracked conveyances (railways), mass production (factories), deep-water vessels (steamships), and more powerful pumps in deep-pit mines. All of these existed prior to the Industrial Revolution, but they multiplied and grew rapidly when coupled with this external source of power. Nearly *every* object that we use in our daily lives—clothing, food, household bric-a-brac—is either directly or indirectly a technological object. Classical technology, such as that built by the Greeks and the Romans, was distinctive and set apart, used by specialists, engineers, or military men. Modern technology, by contrast, brings unprecedented energy, capability, and abundance to nearly everyone in every aspect of modern lives.

In the 19th-century New World of industry, a medical doctor, Jacob Bigelow, in his *Elements of Technology*, was the first to take note of the growth of the useful arts (Bigelow 1829). Among the useful arts that Bigelow devoted chapters to were the "Arts of Conveying Water" (ch. 13), the "Arts of Combining Flexible Fibres" (ch. 15), and the "Arts of Communicating and Modifying Color" (ch. 18). These chapters make it clear that Bigelow's *Elements* was less a theoretical treatise and more an encyclopedia of useful knowledge. A generation later Bigelow and other members of his circle in Cambridge would establish the Massachusetts Institute of Technology, explaining that their intent was to impart "a thorough knowledge of scientific laws and principles" to "the industrial classes" as "a preparation for the labors of the mechanic and manufacturer" (Rogers 1861:28). MIT was the first institution of higher education in the world to devote itself in its very name to technology.

Three national developments, spanning the period from the 1820s to the 1860s and beyond, lie behind this invention of a New Science of technology in America. The first is the opening of the continental interior, through canals, railways, and later steamboats. Early explorers' discovery of the vast expanses of the prairies, the awe-inspiring majesty of the Rocky Mountains, or even the sublime beauties of the Hudson River Valley, furnished American poets and painters with the palette of a new nationalism, defining an original relationship between Nature and Culture (Batteau 1990; Novak 1980). The American nation was the garden of the world, an opportunity for humanity, or at least Europeans, to make a new beginning. It was also, as Leo Marx observes, a Paradise slipping away, a Garden of Eden with a factory whistle heard in the distance (Marx 1964).

For the rising agricultural and industrial classes that placed Andrew Jackson in the presidency, the New Science of technology was about claiming power, political, mechanical, and even social. Leading technological developments during this period stressed the conquest of distance, whether through the Erie Canal (opened in 1825), the first railways (1825), or Morse's telegraph (1844). In the conquest of distance, transportation and communication are always inextricably linked: a lack of means for signaling car movements limited the earliest railways to inefficient one-way traffic. Only when terminals were linked not simply with iron rails but also with copper wires could cars be moved efficiently. Also required is stability of state order, to protect the rails and wires from environmental threats. Complex arrangements such as these, whether in transportation, communication, electrical power grids, industrial supply chains, or even the Internet, distinguishes Technology's world-transforming powers from the numerous local arrangements of ingenuity and usefulness that preceded it.

Closely related to this was the growth of manufactures, beginning with Eli Whitney's invention of the cotton gin, and with his first demonstration of manufacturing rifles with interchangeable parts (1798). By the 1820s, the industrial classes had become a national political force, and in the 1850s, allied with the agricultural classes, pressed Congress for investments in people in the same manner that the government had invested in roads and canals. In 1859 the Congress passed a law, the Morrill Act, establishing the Land Grant Universities, which in its final form provided for instruction,

> without excluding other scientific and classical studies and including military tactic, to teach such branches of learning as are related to agriculture and the mechanic arts, in such manner as the legislatures of the States may respectively prescribe, in order to promote the liberal and practical education of the industrial classes in the several pursuits and professions in life. (U.S.C. § 304)

The joining of science with the useful arts in service of democracy has been a concern of the national government in the United States ever since. It is found today in the billions of dollars in scholarships, fellowships, and research grants dispensed to students, scientists, and universities by the federal government every year.

In the United States, the joining of science, the useful arts, and democracy was a national project, although it was not fully realized until the 20th century. Prior to World War I, most industrial breakthroughs were local affairs, achieved not by men and women with advanced degrees but by practical inventors with little formal training. A machinist who left school at the age of 15, Henry Ford, perfected the techniques of mass production. A pair of bicycle mechanics invented the airplane. Universities began teaching electrical engineering, the technology behind municipal lighting, only after the Edison Electric

Lighting Company illuminated New York City. Only as these local inventions acquired regional and national scope did they become subjects of academic study.

What transformed these local experiments into technological innovation was the building of large-scale, networked systems, whether in railways, electrification, telephony, aviation, industrial supply chains, or automotive transport, both in Europe and the Americas. The development of these networked systems in the 19th and 20th centuries gave rise to engineering as an academic discipline, technology as a popular fascination, and government and corporate laboratories as seedbeds of innovation.

System builders of the past, whether Thomas Edison in New York, Werner von Siemens in Germany, Ferdinand de Lessepps in the Suez, or John D. Rockefeller in the U.S., or even Bill Gates today, impose their vision of uniform order on a plenitude of disparate, local arrangements. To electrify New York City, Edison had to overcome the entrenched interests of other industrialists, the installed base of gas lighting, popular prejudice against a newfangled invention, avaricious local politicians, and the competing standards of other electrical companies. He accomplished this not through any mandate but rather through force of will, consistency of vision, and numerous battles, skirmishes, and advances on the ground as Edison wired up one neighborhood after

Power lines. Electric power transmission is the large-scale transfer of electrical power (or more correctly energy), a process in the delivery of electricity to consumers.

another (Hughes 1983). Likewise, it was insight into software economics and force of will, rather than legislation or technical superiority, that standardized most personal computing from dozens of alternative operating systems to Microsoft's DOS in the early 1980s.

Simultaneous developments in Europe make clear the embeddedness of technology in the national order. European nations industrialized along several unique paths and then used industrial technology to redouble their industrialization, depending on such national conditions as the structure of the class system, the strength and intentions of governments, the political unification of the nation, their geography, and the waning strength of institutions left over from the Medieval era. Britain had the advantage of geography; it was a compact nation with a navigable coastline and rivers, and ample raw materials. A flexible class structure created opportunities for innovation, which industrialists such as Richard Arkwright and R. E. B. Crompton seized. By contrast, in Germany, industrialization would have to wait until political unification (1873), after which Germany emerged as a technological and industrial powerhouse within a generation. Italy, unified at approximately the same time, also undertook aggressive industrialization, at least in its northern regions.[2]

Modern technology is thus an industrial artifact, a product of industrial revolutions in Great Britain, France, Germany, Italy, and later the rest of the world. Earlier civilizations, such as the Greeks and the Romans, can be said to have technologies, insofar as they created manuals and standards for roads, aqueducts, and shipbuilding, and used these systems to coordinate their empires. Likewise, in China under the Qin dynasty, the standardization of weights and measures (221 BCE) was a technological advance, as was the Great Wall of China, the world's largest technological monument.

The nation-transforming projects of building roads and factories and railways in Europe and the Americas were actually projects of large-scale coordination, bringing together multiple languages, customs, and local livelihoods. The Roman roads were large-scale systems, as were the aqueducts built by the Aztecs and the Grand Canal in China, connecting the Yellow and the Yangtze Rivers. That systems such as these are now ubiquitous should not obscure the fact that they are *always* part of projects of imperative coordination spanning multiple cultures, communities, and nations. It is this large-scale project that distinguishes technology from the local ingenuity found in the nontechnological toolkits of simpler societies.

Topologically, a system can be local, point-to-point over significant distance, centralized, or large-scale networked (see figure 2.1). What defines a local system is its embeddedness in a dense social network in ways that further-reaching systems cannot be. Large-scale systems present a set of logistical challenges not found in local or point-to-point

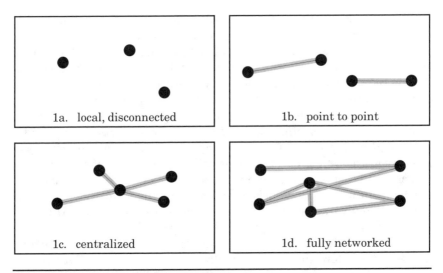

Figure 2.1   Alternative Topologies

systems. A local communication system, an office intercom for example, is quite robust: backups are at hand, and problems can be quickly resolved. A point-to-point system is likewise robust, involving dedicated resources and a stable set of actors. The earliest railways, like the earliest airlines a century later and commercial shipping even today, provided point-to-point services. A pair of destinations is linked by a single line, and goods, information, people, and energy flow back and forth across the line. Telegraphy initially followed the same pattern. In 1853, a resident of Boston, who wished to send a telegram to Albany, would take her message to the Albany and Boston Printing line, located at 77 State Street, whereas if she wanted to send it to Portland, she would have to take it next door to the Boston and Portland Line at 76 State. Other telegraph offices in Boston in this year clearly supplying dedicated lines to single locations included Morse's Northern Telegraph, and the Vermont and Boston Line (Asmann 1980:56).

Municipal systems such as telephones and electric power generation are typically centralized: a central office (for telephones) or power station (for electricity) coordinated traffic back and forth between multiple terminals. This could be quite inefficient; a city such as New York might have four or five incompatible telephone exchanges, each using its own system, and incompatible with the others. A businessman who wished to communicate with all of his customers in the city would have to have four or five telephones on his (secretary's) desk, one for each of the exchanges. In the early years of electrification, when power came from a subscription to central stations such as Edison's Pearl Street Station or one of his competitors, businesses might have a choice of sub-

scribing to a 25, 30, 40, 50, 60, or 66-2/3 cycle system, depending on the
wiring available in their neighborhood (Hughes 1983:128).

For reasons of logistics, centralized systems have limited reach.
The challenges of *continental* integration in the United States and
Europe required the creation of a new operational structure, the networked system. The rise of national railways and railway networks in
the 19th century gave birth to the modern corporate form, with its operational divisions, its functional specializations, and its central staff.
Networked systems, in contrast to even large-scale point-to-point systems, require even greater sophistication to manage flows, balance
capacity with demand, and minimize bottlenecks and traffic jams. The
development of these networked systems, roughly from 1886 (the year
when rail networks in the United States were standardized) to the
1930s, created the modern appreciation of technology.

Human society, of course, has consisted of networks of people and
groups, even before the dawn of civilization. Eric Wolf, in *Europe and
the People Without History* (Wolf 1982), documents the flows of ideas,

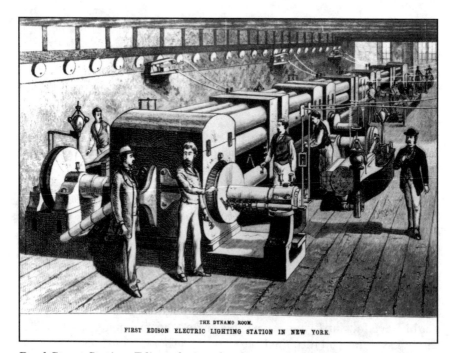

THE DYNAMO ROOM.
FIRST EDISON ELECTRIC LIGHTING STATION IN NEW YORK

Pearl Street Station. Edison designed a direct-current system that was most
efficient for densely populated urban centers and for isolated plants providing
power to a single building. His system was most efficient and economical
within a square mile of the central station. Edison installed his first perma-
nent central station on Pearl Street in lower Manhattan in the 1880s. He
designed special "jumbo" dynamos for the station.

goods, and people in a world without borders. What the Industrial Revolution added to this was first the interposition of mechanical artifacts—iron rails, copper wires, and more recently silicon junctions and glass fibers—into these flows, and the addition of enormous amounts of energy requiring enormous amounts of management. Out of this interposition came Technology as we currently understand it.

In many industries and in many nations one can identify a specific *technological turn* in which industrial development was transformed from a local to a regional or national affair, engineering departments were transformed into laboratories, universities played a more active role, and innovation was mandated either by governments or corporations. In telephony, the technological turn began with Theodore Vail's call in 1907 for "universal service," knitting together multiple local exchanges into seamless, coast-to-coast communication. In the automotive industry, the technological turn began in 1914 with Henry Ford's Highland Park Model T assembly line, continued with his relentless vertical integration of all aspects of engineering, material supply and production, and culminated with the General Motors Futurama, the most widely visited attraction at the 1939 World's Fair. The technological turn marks the end of the lone, self-schooled inventor, whether a Wright or a Ford or a Watt, or a Josephine Cochran, the 1886 inventor of the dishwasher. The technological turn replaced the lone inventor with scientifically trained engineers; the artifacts of wonderful power and social appeal that we associate with technology, whether computers, airliners, or electronic systems, could only be built by teams of engineers working for a government or corporate laboratory, or a university with government and corporate sponsorship.

We might select aviation as exemplifying the technological turn in American industry. For the first third of its history, aviation was either a military option, an entertainment device, or a matter for hobbyists such as Therese Peltier, who, in 1908, was the first woman to pilot an aircraft. The census of 1920 classified pilots as circus performers. As a serious means of transport, it was totally impracticable, even for time-dependent high-value cargo (i.e., airmail). After the First World War, the Army Air Corps began providing airmail service, which was in the early 1920s turned over to private operators, albeit with a generous subsidy. Transport of airmail was through numerous entrepreneurial contracts between pilots, such as Charles Lindbergh, who owned their own airplanes and the postal service. These contracts were for mail transport between pairs of cities: Chicago and St. Louis, for example. This patchwork system struck many as inefficient, although the way out was not clear.

In the early 1930s, the Postmaster General, Walter Folger Brown, rationalized air transport. First, he arranged for an airmail subsidy that would place a premium on an airplane's capacity. Then, after receiving authority from Congress to award postal contracts on a non-competitive basis, Brown brought all of the leading operators together to reorganize the system.

> For two weeks [Brown] cajoled them to swap routes, trade shares of stock in each other, and to do whatever else it took to eliminate duplication, irrationality, and competition—in short, to divide the market to the exclusion of everyone else who had not been invited to the meeting. (Petzinger 1996:10)

The result was that United Airlines had a northern network from Chicago to the West Coast, Transcontinental and Western Air had a central route via St. Louis, American Airlines had the southern routes through Dallas, and Eastern Airlines had north–south routes along the eastern seaboard. Four integrated networks were complete.

In this period, government-promoted rationalization gave birth to passenger aviation and placed aviation in a corporate regime. Passenger aviation was an offshoot of airmail, and in the 1920s and 1930s it had almost no justification other than thrill or status display. Commercial air travel was neither cheaper nor faster nor safer nor more comfortable than rail travel. In the early years, passengers rode among the (subsidized) mail sacks. Only after World War II did passenger aviation become a serious business. The only way it could be promoted was by emphasizing the experience of flight in its modernist aspects. Images of flight drew on modernist themes, altered perspectives, speed, and immediacy. Robert Wohl's *The Spectacle of Flight* (2005) emphasizes the close, symbiotic relationship between flying and motion pictures in the 1930s, both promising new immediacies of experience.

In 1915, the federal government created the National Advisory Committee on Aeronautics. NACA served for 43 years as a government sponsor of the aviation industry's technology, only to be transformed in 1958 to the National Aeronautics and Space Administration to lead the nation in the space race against the Soviet Union. Throughout its entire history, NACA/NASA represented the highest achievements of American technology, even supplying a popular term, "rocket science," for prodigious technological achievement such as space probes and moon landings, and the iconic figure of Wernher von Braun, the German system builder who founded America's space program. The entry of the United States into the First World War placed emphasis on producing aircraft that could fly faster, higher, and farther. The beginnings of commercial aviation in the airmail and later passenger lines of the 1920s created new opportunities for expression of modernist themes, experienced by the public with the cachet of "airmail" and celebrity air travel.

NACA's first breakthrough contribution to aviation came in the late 1920s with the invention of the NACA cowling, a streamlined shroud covering the cylinder heads, and thus substantially reducing drag and improving efficiency. Prior to this, the fully exposed cylinder heads of radial airplane engines, such as the Wright Whirlwind J-5C on Lindbergh's *Spirit of St. Louis*, were a major source of drag. Efficiencies provided by this cowling and other inventions such as retractable landing gear turned aviation into a commercially viable proposition. The safe, easy, and affordable air travel that many enjoy today was made possible by this exuberant history of experimentation and development.

These themes of efficiency and affordability were fully realized in the DC-3, the first commercially successful airliner, in 1935. The DC-3, with its Art Deco profile and enhanced cabin capacity (even allowing for sleeper berths, thus removing the image of heavier-than-air flight from the wild blue yonder and placing it firmly in a domestic space[3]) was the beginning of a decades-long marriage of form, function, and meaning in civil aviation. In the early 1930s, the integration of disparate "lines" such as the Chicago–St. Louis line into national networks— the large-scale systems—was accomplished not through any natural market evolution but through the vision, inducements, and browbeating of a monopsonistic customer.

A group of French anthropologists, Alain Gras, Caroline Moricot, Sophie L. Poirot-Delpech, and Victor Scardigli, likewise have described the growth and representational consequences of the creation in Europe of *un macrosystème*, civil air transport (Gras *et al.* 1994:17). Their analysis of system development is consistent with and reinforced by my conclusions: the importance of system building, the role of the state, and the role of international bodies in harmonizing technical relations among nations, standardizing the actions of actors, synthesizing views on how to develop air transport, and centralizing the conditions for its further development.

After an early start with daredevils, bicycle mechanics, and lone eagles such as Louis Bleriot and Harriet Quimby, aviation adopted technology and the technological ideal as a critical element of its growth. Before this technological turn, the science of aeronautics and the experiments of aviators proceeded along two unconnected tracks (Anderson 2002). By the end of the 1920s these tracks converged, and by the end of the 1930s, aviation was firmly established as one of the most technologically advanced industries in the United States, and universities were creating departments of aeronautical engineering. This was far from the case even when Lindbergh flew.

REPRESENTING TECHNOLOGY

Large-scale systems such as these require uniform standards, in a manner that local tools do not. Prior to 1886, railways in the United

States had multiple track gauges (the spacing between the rails). Railways in the northern states tended to use standard gauge (56.5 inches), whereas in the southern states broader gauges were used. After the civil war, this break in gauge, as cars traveled between North and South, was a major economic inefficiency; cars had to be removed from the tracks, and the spacing of their wheels reset, an operation that might require an entire day. Similarly, prior to Henry Ford's assembly line, making automobiles was a craft production, and each master craftsman had his own tools and gauges with which he measured and finished parts. The result of this was that two teams' inches might be fractionally different. Whereas this discrepancy might be acceptable for one-off craft production, or for interchangeable parts made by a single master such as Eli Whitney, it is not acceptable for large-scale mass production where parts and producers must be fully interchangeable. Henry Ford took away the masters' gauges and substituted his own standard tools. Likewise, the building of municipal electric grids required agreement on what type of current would be supplied: direct or alternating (which Edison opposed), single-phase or polyphase, 50 or 60 cycles, and 110, 220, or some other voltage. Political institutions and forceful personalities, and not technical superiority, resolved this "battle of the systems" (Hughes 1983:122–128ff.).

An engineering standard is a unique cultural and linguistic artifact, no less essential to a technological society than a system of laws is to a civil society. Like civil laws, engineering standards represent a rational codification following an irrational process in which contending parties argue, sometimes forcefully, for their own view. Different devices—incandescent lamps, induction motors, arc lighting, transformers—run best on different types of current. If a company has its major investments in incandescent lamps, or induction motors, or transformers, etc., it is going to favor one form over another and argue forcefully that theirs should be the universal standard. Engineers are familiar with "standards wars," in which contending parties slug it out, perhaps in the marketplace, perhaps in universities, perhaps in standards bodies such as the International Organization for Standardization (ISO), for the superiority of their local version. These "technological dramas" (Pfaffenberger 1992b) have all of the tension, uncertainty, and theatricality of contests in the larger society. Yet once they are settled, and the standards are published, they have the force of law and the authority of Science behind them.

It is the basis in standards that distinguishes technological artifacts—with the learning and the power they represent and the social compromises that they embody—from local tools such as shovels or digging sticks.[4] This also firmly embeds technology in linguistic constructions, one of the guiding questions I raised in the previous chapter. These local tools are certainly quite useful, but not quite technological.

A standard is a collective representation, no less so than other cultural representations. Like other collective representations, a standard embodies a great condensation of meaning, albeit accessible only to those who are aware of its creation. For example, a standard such as 60-cycle, 110 volt polyphase alternating current may seem coldly rational and scientific, until one acquires knowledge of the struggles between Thomas Alva Edison and Charles Proteus Steinmetz, the false starts in building municipal lighting systems, the political conflicts in many cities between gaslight interests and the upstart electricians, and the deaths of nameless members of the industrial classes who were learning the power of lightning-by-wire. The relationships and conflicting social values that make up a standard are condensed within these engineering standards, which acquire forceful meaning through dramatic conflict and are lent nobility by the sacrifices of failed efforts, unsuccessful inventions, and wasted years that preceded them. A tool does not become technological until it acquires this representation, graced by a learned elite.

Technological dramas are thus not just superficial decorations for the ongoing development of technology; they are constitutive of that history. These dramas enroll broad communities of engineers, manufacturers, users, and maintainers in an agreed-upon configuration that then becomes a platform for further development. Without the agreed-upon configuration, electrification would have been hit-or-miss, disconnected, and not socially transformative. The resolution of the struggle settles the matter not just for the protagonists but also for an audience constituted by the broader society.

The wonder-working power of technology lies ultimately not in its packaging of millions of ergs, watts, or joules of physical force and energy but rather in its packaging of struggles, compromises, and values representing the millions of people in society. Physical force and energy are nothing but earthquakes and thunderstorms until they are tamed by society. Taming these forces and making them socially beneficial is always a collective project, albeit obscured by heroic myths of lone inventors (Prometheus, Franklin, Edison, Ford, Jobs) and the sleek packaging of the final result.

There is a close relationship between technology, as we understand it today, and Modernism. Anyone can see this relationship in the styling of technologies and their design. Popular assumptions about technology are infused with modernist assumptions and aesthetics, to the extent that if a useful device does not conform to these standards, its status as "technology" is questionable. Modernist aesthetics inform technology not only in its form factor (the shape of the external package) but also in its architecture (integration rather than end-user assemblage) and its aesthetic priorities (hiding functionality behind smooth surfaces). The contemporary cell phone, which packages large

amounts of functionality into a small, rounded package, is an example
of modernist technological design.

From the earliest days of the Industrial Revolution in the 18th century, technology has infused Modernism, and Modernism has infused
technology, although it would not be named as such for at least another
century. The earliest prophets of Modernism, including Goethe and
Marx, appreciated the power of industrial development and the world-
altering capabilities that its tools gave to humanity. Marx and Engels,
in their *Communist Manifesto*, took note of world transformations:

> The bourgeoisie, by the rapid improvement of all instruments of pro-
> duction, by the immensely facilitated means of communication, draws
> all, even the most barbarian, nations into civilization. The cheap pric-
> es of commodities are the heavy artillery with which it forces the bar-
> barians' intensely obstinate hatred of foreigners to capitulate. . . .
>
> The bourgeoisie, during its rule of scarce one hundred years, has
> created more massive and more colossal productive forces than
> have all preceding generations together. Subjection of nature's forc-
> es to man, machinery, application of chemistry to industry and
> agriculture, steam navigation, railways, electric telegraphs, clear-
> ing of whole continents for cultivation, canalization or rivers, whole
> populations conjured out of the ground—what earlier century had
> even a presentiment that such productive forces slumbered in the
> lap of social labor? (Marx and Engels, 1992[1847]:18f)

Likewise, the German poet Goethe, in *Faust*, his narrative of a
medieval philosopher-magician who sold his soul to the Devil in return
for extraordinary powers, understood the world transformations that
could result:

> Green are the meadows, fertile; and in mirth,
> Both men and herds live on *this newest earth,*
> Settled along the edges of a hill
> Raised by the masses' bold, industrial will.
> A veritable paradise inside,
> Then let the dames be licked by the raging tide,
> And as it gnaws, to rush in with full force,
> Communal will fills gaps and checks its course.
> This is the highest wisdom that I own,
> The best that mankind ever knew;
> Freedom and life are earned by those alone
> *Who conquer them each day anew.* (Goethe, 1963:467f, italics supplied)

As a cultural movement, marking a tectonic shift in values and
objects of social attachment, Modernism undercuts classical concepts of
order and hierarchy, celebrating the individual, innovation, and human
freedom. As an aesthetic movement in painting, literature, and music,
Modernism has given perfection of form and sensuous appeal to the dis-
tortions of perspective, power, and sociality that the tools of industrial-

ization created. The industrial aesthetic (smooth surfaces marked with only a few focal points, hiding of functionality so that transformations of force or matter seem yet more miraculous) is part of the image of technology. Modernism infuses these awesome systems with an aura of mystery and sublimity, creating a set of expectations that today define technology. These images and expectations raise "technology" above mere usefulness, allowing technology to give meaning to lives in ways that shovels and digging sticks never could. Many technological objects, especially those with stupendous powers, are objects of divine inspiration, as David Noble describes in *The Religion of Technology* (1997). When we speak of some object as "high tech," we are referring as much to its sleek lines, smooth surfaces, and magical expectations as we are to its engineered functionality.

In the first decades of the 20th century, the technological and Modernist distortion of perspective and space accelerated. Humans began to fly, spanning continents and oceans within a matter of hours. Edison electrified New York, turning night into day. Hearing distant voices became not a hallucination but a normal business practice. Teenage couples zoomed away from parental supervision at death-defying speeds in the Model T. F. T. Marinetti's *Futurist Manifesto* (1909) smashed staid convention as it celebrated the "joy of mechanical force." The creative destruction of time, distance, perspective, and social relationships was underway.

This marriage of technology and Modernism received an official blessing with the heroic technological projects of the 1930s. The New Deal plunged the federal government wholeheartedly into technological developments, whether with the Hoover Dam, the Bonneville Power Administration, the Golden Gate Bridge, or major support for the aircraft industry. Each of these was a scale-defying use of the new tools, materials, and human wants that technology supplied. The architect of one of these projects, David Lilienthal (1944), chairman of the Tennessee Valley Authority, characterized their efforts as "Democracy on the march." All of these projects were portrayed as bringing the benefits of industrialization—employment, electrical appliances, easy mobility—to backward regions: Democracy through Technology.

It is this assemblage of mechanical devices, social values, industrial production, monumental structures, and governmental authority into a large, self-propelling machine that justifies speaking of the "invention" of technology, a machine that hurtles forward into the 21st century as the "military-industrial complex." In the 19th century, the various components of this assemblage lay in separate parts-bins: some laws of physics stored in universities, public authority latent in the Constitution, and an untapped workforce resting idly on front porches along the Tennessee River. Systems builders, the prime movers of technology, whether Edison in the 1880s, Ford in the 1910s, Lilienthal in the 1930s, or von Braun in the 1950s, brought these pieces together to create regional and national networks of power, movement, and social transformation.

By the end of World War II, Technology had displaced Providence as an object of faith in American culture. Technology became democratized. Technology was no longer the specialization of learned Brahmins such as Bigelow, nor was it the awesome but exclusive possession of the engineers such as Lilienthal. As the economy grew in the postwar years, Everyman could have technological devices such as radios and automobiles in his living room or his driveway. Everywoman could have time-saving and convenient devices such as washing machines and refrigerators in her home. Everybusinessman could have ingenious devices such as calculating machines and dictaphones in his office or factory. And finally, Everypresident could imagine heroically leading the Nation to new, monumental, technological achievements. Comprehending these stupendous achievements poses new challenges for anthropology, the science of humanity.

Notes

[1] This phrase "island communities," from Robert Wiebe's *The Search for Order* (Wiebe 1967), nicely captures the loose-knit character of American society prior to the development of rail networks.

[2] One of the intriguing puzzles of technology is why the originators of a new technology often fail to capitalize on it. The Chinese invented gunpowder and the compass but did not allow them to disturb their social order. Xerox Palo Alto Research Center invented the graphical user interface (GUI); the adoption of the GUI by Apple turned it into a revolutionary technology. The best that can probably be said here is that technical cleverness and economic transformation are completely separate matters.

[3] Factually, it precariously perched a domestic space in the wild blue yonder.

[4] After this was written, one of my students, Felicia George, gave me an advertisement for a treatise by Frederick Taylor on "scientific shoveling," in which he argued for aligning and standardizing shovels based on the material being shoveled.

Chapter Three

An Endless Frontier

Victory in battle, in the words of Confederate General Nathaniel Bedford Forest, goes to the army that arrives on the battlefield "the firstest with the mostest." Massive use of force, smartly deployed, is the difference between winning and losing. The Civil War was the first modern conflict in which the tools of industry—railway, telegraph— enabled armies to get to the battlefield the firstest. In the Northern states, the Industrial Revolution supplied the Union Army with the mostest, outproducing the agrarian South. President Abraham Lincoln had to work his way through five generals before he found the right one, Ulysses S. Grant, who knew how to use industrial methods and resources to grind down and annihilate the Confederate Army.

The Industrial Revolution of the 19th century industrialized and technologized warfare. Ever since then, technological breakthroughs have led to military advantages, and wartime mobilizations have accelerated technological innovation. For the first third of the 20th century, this relationship between the military and the scientists and technologists was a maverick affair: oddball army officers, such as Billy Mitchell, experimented with newfangled devices, such as airplanes and rockets, and a few longhair scientists were happy to receive this attention. For the most part, though, scientists and military men regarded each other with mutual suspicion.

In this chapter, I examine how governments and corporations have supported technological innovation as a means of achieving military, diplomatic, or competitive advantage. In particular, I focus on the distance between discovery and invention, on the one hand, and the harnessing of discovery and invention through innovation to socially accepted goals, on the other. I examine how government and corporate laboratories have tried to shorten this distance, not out of a concern for the betterment of humanity, but in order to gain advantage over diplomatic or commercial rivals. Ironically, enlisting technology in the service of corporate or national objectives has widened the gaps between invention and the broad application of new technologies, as corporations and governments attempt to monopolize technological advantages.

THE GATHERING STORM

Wars accelerate technology—the more terrible the war, the greater the acceleration. In the 1930s, as war with Germany approached, President Franklin Roosevelt understood that as the war would match industry with industry, it would also match technology with technology. In the Battle of Britain, fought in the skies over England in the summer of 1940, a new technology, radar, gave the much smaller Royal Air Force an advantage over the numerically superior German Luftwaffe. Although the early science of electromagnetic radiation was discovered in Germany by Heinrich Hertz, and both Germany and Britain were developing radar systems in the 1930s, only British Air Chief Marshall Hugh Dowding sufficiently understood how to combine this new way of seeing with new ways of tracking, reporting, and mobilizing squadrons of aeroplanes in order to encounter the enemy in the skies the firstest with the mostest.

This contrast, between the discovery or invention of a new device or technique or natural phenomenon (Hertz) and its successful deployment (Dowding), is at the heart of the problem of technological innovation. Ever since World War II made technology the key, not only to the advance of civilizations, but also to the survival of civilization, governments and industries have searched for the means by which they could achieve technological superiority. More accurately, they have sought to find the best means for using technology to achieve national goals, whether those goals are the modernization of their economy, the eradication of tropical diseases, or the provision of an adequate livelihood for their citizens.

The gap between invention and innovation had been a concern among anthropologists, rural sociologists, and communication researchers for most of the 20th century. This interest comes out of a modernist conceit, wonderment at why farmers or village-dwellers or tribalists or educators did not *immediately!*[1] adopt some putatively superior method or artifact of irrigation, or hybrid corn, or pedagogical technique. From today's perspective, the wonderment is not so much why innovations are not immediately adopted, as to why anyone might assume that they should be. Only a perspective that axiomatically equates technology with reason and progress and, therefore, any lag in its adoption with stupidity or recalcitrance, would make such an assumption.

The lag between the existence of a New Thing and its widespread adoption had been noted at least since the French sociologist Gabriel Tarde concluded that imitation is a distinctive social fact (1903), but it did not become an urgent problem until war clouds began to gather in Europe in the late 1930s. In January 1941, Waldemar Kaempffert, a science writer with the *New York Times*, noted that military application was a great driver of technological innovation, not invention, and that the exigencies of wartime frequently speeded innovations into widespread application. Kaempffert and others used William F. Ogburn's concept of "cultural lag" (Kaempffert 1941; Ogburn 1922) to explain why society at large did not keep up with new ideas and artifacts coming out of university laboratories at the time.

This was the problem that motivated Vannevar Bush in the summer of 1940 to propose creating the National Defense Research Council (NDRC) within the War Department. Bush was a former university administrator, inventor of one of the earliest analog computers, a member of the National Advisory Committee on Aeronautics, an entrepreneur, and a cofounder of the electronics manufacturer Raytheon. He was the president of the Carnegie Institute of Washington, one of the private philanthropies that exemplified pre-War support for scientific research. Bush had a Yankee ingenuity and stubbornness that would not wait for new weapons to "diffuse" through a hidebound military establishment. He knew that he was smarter than most military men, and only when required by diplomacy did he hide the fact. For the next five years, Bush and a small circle of colleagues dominated America's scientific elite, establishing a set of research priorities and relationships that continue even today.

In the summer of 1940, on the eve of the Battle of Britain, President Roosevelt, on Bush's recommendation, chartered the NDRC, which in less than a year evolved into the congressionally authorized Office of Scientific Research and Development (OSRD). This was a first-of-its-kind partnership among government, industry, and elite universities—the beginnings of the military-industrial complex that Eisenhower warned about 20 years later. By the end of the war, OSRD had dispensed billions of dollars in contracts, primarily to a favored few institutions and corporations, many of which had close, prewar ties to Bush. Like Edison, and like Lilienthal, Vannevar Bush was a system builder, only his system was less an assemblage of electromechanical devices than it was a system of expertise for producing electromechanical devices. His lean, angular, close-cropped looks and ever-present pipe established the archetype for government-scientist-administrator for generations to come.

Vannevar Bush characterized World War II as "a race between techniques" (Zachary 1997:169). At the beginning of the war, the Germans had both quantitative and qualitative superiority in aircraft, submarines, field guns, and tanks. They were able to use these advantages to overrun Europe, cut off England from receiving vital supplies, and threaten American shipping. The Germans understood radar but, as noted, never deployed it successfully. The OSRD was responsible for the several inventions, rapidly fielded, that turned the tide against Germany in 1943 and 1944. Among these were microwave radar for hunting submarines and proximity fuzes for improving antiaircraft fire. Also developed under OSRD was a technology, not decisive in the war, but that was pivotal in the uneasy peace that followed: nuclear power.

The Second World War and its dramatic conclusion, more than any other event or epoch, established an equivalence of "science," "engineering," and "technology" in the public mind, eventually leading to an awkward locution of "technoscience" in academic literature (Haraway 1997; Law 2002). Conceptually, these are quite distinct: science is about discovery, engineering is about making things work, and technology is about things that work. This alliance—the utilitarian expectations for science, the enlistment of arcane natural phenomena to make things work, and growing public acclaim for the engineers who make them work—represents a profound cultural shift, a conceptual and institutional transformation that reverberates well into the 21st century.

LA PENSÉE SCIENTIFIQUE

In July 1945, as World War II was drawing to a close, Bush submitted (at Roosevelt's request) a report that has served for the decades since as the charter for federal support of science. This report, *Science, The Endless Frontier*, lays out a program of federal support for basic research in the natural sciences, in medicine, and in the training of new scientists. Drawing on romantic imagery of America's "pioneering spirit," it suggests that the cultivation of science and scientists both is of national importance and is firmly grounded in the nation's heritage:

> The pioneer spirit is still vigorous within this nation. Science offers a largely unexplored hinterland for the pioneer who has the tools at hand. The rewards of such exploration, both for the Nation and for the individual are great. Scientific progress is one essential key to our security as a nation, to our better health, to more jobs, to a higher standard of living, and to our cultural progress. (Bush 1945: letter of transmittal)

In the view of *Endless Frontier,* which has been authoritative for most of the 20th century, technological innovation rests on a foundation of applied science, which in turn rests on a foundation of basic research. Most of the research carried out by government agencies and private corporations is applied research—research aimed at solving a particular problem, whether the conductivity of electrical transmission networks or the strength of materials used in aircraft. The Applied Scientists, in this view, make withdrawals from a storehouse of knowledge stocked by the Basic Scientists and work them up into useful New Things. Various estimates of the length of time required to turn Basic Science into commercial products range from 10 to 30 years. For example, the principles of electrostatic imaging were discovered by Hungarian physicist Paul Selenyi in the 1920s. These principles would not have an application before 1938, when a patent attorney, Chester Carlson, annoyed by the inefficiencies of carbon paper, had the idea of using Selenyi's discoveries to make copies. Carlson was assisted by the Battelle Memorial Institute, which agreed to share royalties, and developed the process, later sold to the Haloid Corporation. Haloid was renamed Xerox and sold its first commercially successful copier in 1959.

Breakthroughs such as xerography are the stuff of industrial legends, so ubiquitous in recent decades that we fail to appreciate how revolutionary they were only a few generations ago. Other breakthroughs of the postwar era include the transistor (Bell Labs, 1947), the graphical user interface (Xerox Palo Alto Research Center, 1981), liquid crystal flat-screen displays (IBM Watson Research Center, 1980s), the Local Area Network (also PARC, 1973), and, of course, the Internet. The technological triumphs of World War II prompted numerous corporations to begin paying attention to technology both for purposes of dominating their existing markets and for creating new markets.

Leading corporate laboratories had been founded earlier in the 20th century as part of the technological turn in key industries, frequently led by the system building described in chapter 2. In telecommunications, Theodore Vail's 1907 call for "universal service" stitching together disparate telephone exchanges into one system, along with the 1915 founding of Bell Labs, marked a technological turn; in aviation, the growing success of the National Advisory Committee on Aeronautics, founded in 1915, was a similar technological turn. By the 1950s, many large manufacturing corporations were transforming their engineering departments into laboratories, creating research centers that resembled university campuses and seeking the next breakthrough, the next Carlson or Selenyi who would propel them to industrial dominance.

A vast cultural experiment in the re-arrangement of *technê* and *logos* was underway, an experiment that ran from the 1940s until the 1980s, when competitive pressure from globalization forced many companies to downsize or offshore their research operations. This experiment was based on a linear model of the relationship between Basic Science, Applied Science, engineering development, and commercialization, a model that has been surprisingly immune to repeated empirical refutation. In this experiment, companies sought, usually without success, to build durable and productive relationships among scientists and engineers on the one hand, and their factories, their markets, and their customers, on the other. Was it better to have one's own pipeline of basic research to applied research to product development, and stuff the front end of the pipeline with Basic Science coming out of an ivory tower? This was the model for Bell Laboratories, and it worked as long as AT&T was a telecommunications monopoly. Other companies, such as Xerox, lacking monopoly position, were often unable to gain benefits from their own breakthroughs in Basic Science. Or, was it better to place Research and Development (R&D) close to manufacturing and capitalize on synergies between the two? Some of IBM's greatest breakthroughs came not from its world-class Thomas B. Watson laboratories but from R&D activities buried in operating divisions. Should industries cooperate in research consortia, such as Sematech or the National Center for Manufacturing Sciences, and should the federal government underwrite these consortia? Should we nurture new ideas coming from the longhaired boys in the lab, from the skilled hands in the factory, or from the customers in the field? Hundreds of books and tens of thousands of business-press articles have offered a variety of answers to these questions (Mowery and Rosenberg 1998).

This intense activity, in addition to creating such marvels as computers and cellular telephones and new vaccines, created an expectation that there was no problem beyond the reach of technological solution. As this proposition has been debated ever since, one should note that it contains within it a strong element of magical thinking. In the 1950s and 1960s American society, probably more so than any other society in the world, began to associate with technology a magical sense of mystery and unlimited capability (Gell 1988). Expectations of this sort, which have taken wing and diffused all over the globe, have created an unavoidable tension between technology's supposed capabilities and its actual, disappointing benefits.

The large institutional investments by both governments and corporations in technology gave rise to a discipline of "technology assessment," an effort to understand the benefits and prognosticate the payoffs of these investments. Technology assessment examines such issues as the costs and feasibility of fuel-efficient cars, the benefits of energy conservation measures, or the prospects for extending the Internet and computer literacy to households across the nation. Although technology assessment presents itself as a technocratic discipline, in fact it has strong political overtones, a fact that led to the shutdown of the Congressional Office of Technology Assessment following the Republican takeover of the House of Representatives in 1994.[1]

Over the past century, the linear model has been sufficiently iconic and durable that we must conclude that it is primarily a totemic representation, impervious to empirical verification. The linear model appears in many different forms and contexts: in the differential prestige and expectations of "basic" and "applied" subdisciplines of several fields, in the budgeting policies of many agencies sponsoring scientific research, and in an array of professional societies. The linear model has been repeatedly criticized as simplistic and misleading. The linear model assumes that scientific discoveries are translated automatically into technological developments, which are translated automatically into new products and new markets. In fact, as many new companies have demonstrated (Amazon, Dell, Federal Express), product innovation, industrial innovation, market innovation, and service innovation are not dependent on scientific discoveries. These companies have achieved breakthroughs by making creative uses of here-and-now technologies. Yet, the linear model is practically cast in stone in the mission of the National Science Foundation and the structures of many corporations and universities. Like all totemic representations, as Claude Lévi-Strauss describes in *La Pensée Sauvage*, (Lévi-Strauss 1963), the linear model establishes an array of oppositions and equivalences (see figure 3.1), thus practically dictating that the nobility of Basic Scientists should have a commanding position. As a totemic object, the linear model is a key symbol (Ortner 1973) of science and technology in American culture.

Basic		Pure		Visionary		Noble
is to	*as*	*is to*	*as*	*is to*	*as*	*is to*
Applied		Impure		Practical		Ignoble

Figure 3.1   Linear Totem (compare Lévi-Strauss 1962:141)

Totemic equivalences, such as these, in tribal societies associate clan members with clan totems; thus, members of the Eagle clan speak of themselves as eagles and members of the Bear clan as bears. A sense of identity and exclusivity is derived from these characterizations, marked by terms such as "clannish" or "tribal." Even today, ongoing debates over the boundaries between "basic" and "applied" research are less about epistemology and more about budgetary entitlements and the identity politics of the scientific community. Debates within different disciplines over research priorities and methodologies (quantitative surveys v. qualitative description, for example) pivot less on a developed understanding of methodological alternatives than on views of "thick description" or "multivariate regression" as totemic objects.

INNOVATION AND ITS DISCONTENTS

Commercial successes such as Xerox, and numerous successful and failed efforts to innovate in the 1930s, 1940s, and 1950s, created a sociotechnical perspective on technology. This sociotechnical perspective, summarized here, is authoritatively presented in Everett Rogers' book *Diffusion of Innovations* (2003). Building on earlier archaeological and anthropological observations that observed specific artifacts and techniques geographically spreading among communities, diffusion theories tracked and sought to explain the eager or slow adoption of a large variety of New Things, from contraceptive practices to typewriter keyboards to the now-ubiquitous computer mouse. Innovation rates were explained in terms of multiple factors: social networks, characteristics of the artifact or technique (scalability, trialability, observability, and other characteristics), mass media, opinion leaders, and cultural compatibility or lack thereof. A substantial amount of research established a consistent pattern, a logistic or "sigmoid" (S-shaped) curve, in which early adopters slowly build a critical mass or critical market for the innovation. Once this "inflection point" is reached, adoption accelerates until it reaches some saturation point.

As Rogers notes, diffusion research has several limitations or blind spots. Diffusion research tends to have a pro-innovation bias, a focus on the individual (leading to blaming those who "fail" to adopt). Further, diffusion research tends to overlook second- and third-order consequences of innovations. Diffusion research also has little to say about innovation's tendency to "increase socioeconomic inequality [which] can occur in any system," and which is especially noted "in Third World nations" (Rogers 1995:125). These shortcomings have only

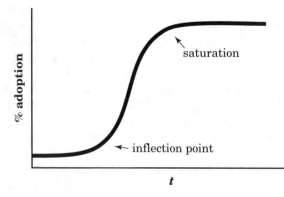

Figure 3.2 Logistic Curve

minor consequences when one's frame of reference is a rural county in Iowa, or the diffusion of a small-scale object, such as a sewing machine, or a pedagogical technique. When one is considering the global diffusion of large-scale systems, whether transportation networks or industrial supply chains, there are several issues (noted below) that diffusion models do not adequately explain.

Another limitation of diffusion research is that it does not make a clear distinction between the "natural" diffusion of innovations, such as improved farming practices in the American Midwest spreading from farmer to farmer, and imperatively coordinated impositions, such as the adoption of new computer systems by a corporate IT department. Since the 1980s, numerous corporations have attempted to improve their processes by introducing new technology; this effort, under the banner of "reengineering," was frequently resisted by rank-and-file workers who worried what the new technology would do to their jobs.

The distinction between this directed diffusion and natural diffusion is actually more subtle than it may appear on the surface. On the one hand, Agricultural Extension Workers, sponsored by the U.S. Department of Agriculture, have been working with farmers for decades to improve agricultural practices; their main tools have been persuasion and demonstration. Significant advances in farming techniques in the 20th century can be credited to these Extension Workers. On the other hand, directed innovations, such as new computer systems, may be resisted for years (or even longer) as corporate employees cling to tried-and-true methods, even to the point of maintaining bootleg pencil-and-paper systems with which they get the job done when IT is not looking. The factual adoption of such innovations, managerial intentions and theory to the contrary notwithstanding, is just as dependent on personal ties, demonstrations by early adopters, the advocacy of champions, and cajoling by supervisors. The importance of this distinction is thus not so much in the actual process of diffusion as it is in the policies one adopts and the resources one commits in the pursuit of innovation.

When innovations are imposed from above, they are at times resisted, passively or actively. Those who resist technology are given the name "Luddites," after an 18th century movement in England protesting the introduction of water-powered looms. The epithet "Luddite" implies an unthinking resistance to technology. The ambivalence with which some embrace or reject it is expressive of deep-seated ambivalences toward technology. In fact, the 18th-century frame-weavers, who secretly identified themselves with "Ned Ludd," who were smashing the power looms were not so much opposed to the technology *per se,* as they were to the poorer quality of industrially manufactured cloth (Thompson 1963). Likewise, many who today are tarred as "Luddites" resist innovation, not for its novelty, but also for the degraded working conditions, increased surveillance, or erosion of social ties that technology often brings (Braverman 1975; Emery and Trist 1965; Susman 1976; Trist, Susman, and Brown 1977). Technologies' advantages are always accrued in an uneven manner, although this fact is easily overlooked when viewed from above.

One class of innovation-from-above, military technology, has established a link between state power and engineering innovation that dates back several millennia. Although war has long been a stimulus for economic and industrial innovation, its dynamics of secrecy, destructive force, and cultural regimentation are not addressed in the diffusion literature. When political leaders are tempted to treat technology as a spectacle, such as Kennedy's desire to send men to the moon, or Nixon's desire for a cure for cancer, this political calculus frequently trumps economic or scientific or social objectives. Diffusion theory is silent in the face of the great or terrible things that governments do with technology.

A related distinction that the innovation literature glosses over is that between useful artifacts and techniques in general, and technologies in particular. Technologies are defined by their usefulness and by their scale, their connectedness, their aesthetics, their condensation of extensive social relationships, and their embodiment of authoritative standards. Thus, when a new technology arrives, the recipients are negotiating its usefulness and its scale, and a redefinition of sociality and identity that it implies. For example, the adoption of patented varieties of hybrid corn is not simply a matter of a farmer improving his yield; it also requires new farming practices and new relationships with the seed suppliers, his neighbors, and the markets for his products. The diffusion of innovation and the diffusion of technology are two separate and not fully commensurable problems.

A not very original observation is that original inventions or discoveries are few and far between. Most innovations creating dramatic performance breakthroughs are a matter of recombining familiar artifacts for unfamiliar purposes. Hertz's discovery of electromagnetic radiation was original, as was the invention of Morse's telegraphic code, but Marconi's recombination of these into the first radio transmission, and the later addition of amplitude modulation to carry voices over AM radio, was pure recombination. Inventors and innovators are often spoken of in the same breath, when in fact the two activities are quite different. Invention is about *disconnection*—breaking the surly bonds of familiar habit, or plucking an odd phenomenon out of its original context and making new sense of it. Innovation, on the other hand, is about *reconnection*—reknitting social bonds that were disrupted by some new artifact, industry, or social group. Proof of this is in the null set of "unsuccessful innovations"—new ways of doing things that never took root. Innovation, by definition, is successful; this is what distinguishes it from invention or novelty.

In the 1970s, the personal computer was a hobbyist's device, a microprocessor plugged into a printed circuit board along with several switches, of no value except impressing the other social misfits at the San Francisco Bay Area Home Brew Computer Club. *Real* computing still belonged to IBM's big iron. In 1975, the first personal computer, the Altair 8800, was sold in kit form by a company called MITS. The evolution of the personal computer, as described in Robert Cringely's *Accidental Empires* (Cringely 1992), did not evolve out of downsizing the *real* computer but out of experimentation by hobbyists and nerds with radios and other electronic devices, promoted by Radio Shack's TRS-80. Only in the 1980s, when computers came bundled with useful software including spreadsheets (the original "killer app"), and in the 1990s, when microprocessors were combined with graphical user interfaces did personal computing reach the inflection point, turning from a neat-to-have object for impressing the guys into a gotta-have tool for impressing the boss. Similar stories of experiments in combining configurations, markets, applications, and contexts before hitting the winning result can be told about every successful innovation.

But just what is recombined? Demonstrably, it was not just a new way of arranging wires and circuits that made AM radio a durable innovation, but new groups, new functions, new topologies, and new expectations also played a role. Such recombinations produce new relationships among users and between users and artifacts. These relationships are among the most far-reaching consequences of the innovation. Radio embraces not just transmitters and receivers; it encompasses and reaches out to corporate interests, advertisers, regulators, entertainers, journalists, and, yes, audiences. In stitching together broadcasters and national audiences, radio created national markets for other new products. Likewise, Air Chief Marshall Dowding's successful deployment of radar embraced not just antennae but also an elaborate network of communication, evaluation in a Filter Room, where reports from 21 radar stations were sifted, and an Operations Room, which dispatched squadrons of fighters from more than 50 airfields. In short, Britain's decisive advantage was not simply the device (radar), but it was, in Dowding's words, "science thoughtfully applied to operational requirements" (Townsend 1970:173).

This view, of technology as comprised of more than simply artifacts, and embracing values far more diverse than simple utility, has been developed by a group of European and American sociologists, anthropologists, and historians, under the heading of actor network theory (ANT), a viewpoint that I will develop further in chapter 7. The actor network approach views technologies as emergent webs of social groups, artifacts, and problems. Its emphasis is not on the artifact itself (or any other single element in the system), but on the properties of the network as a whole. In these networks, the artifacts and the problems as well as the social groups are seen as restless, searching for connections with other groups, artifacts, and problems, until the network is stabilized. In the view of actor network theory, it is not the superiority of any one artifact or standard but the characteristics of the entire network that determine its trajectory. In automotive transport, it was the self-reinforcing combination of rising incomes, improved roads, and the mass-production of an easy-to-use device (the automobile itself) that propelled America into the automotive age in the 1920s. Occasional setbacks, such as rising accident rates as roads and speeds were improved, have to be either incorporated into the system or defended against.

Forward salients are the forays of the system into new territory, where once again it must establish and stabilize itself. The development of macadam asphalt highways, for example, invented by the Scottish engineer John Loudon MacAdam, created the smooth surfaces that were needed for automobiles to attain their speed potential. These militaristic metaphors—forward and reverse salients—suggest that one factor that establishes the final result is which system can establish itself as the firstest with the mostest. This is certainly how the Edison interests won the "battle of the currents," how the VHS format for videocassettes overcame the technically superior Betamax, and how Microsoft came to dominate the software market for personal computers.

The close link between state and corporate power and technology in the 20th century is evident in the enlistment of government sponsorship in the service of one technological alternative or another as well as in a new class of systems, "monumental technology," that achieved its highest expression in the Apollo and Soyuz space programs. As conceptualized by Carolyn Psenka (2008), monumental technologies are those achievements of such awesome size and complexity that, in comparison to their overwhelming costs and their real but not overwhelmingly practical benefits, they must be seen primarily as displays of state power: "Look on my works, ye Mighty, and Despair!" Certainly, the Apollo program qualifies here: the monumentality of the achievement, the sublime experience of the launch, and the epoch-shattering experience of watching—"live"—men walking on the moon were breathtaking for billions of people on Earth. Other monumental technologies in this class if not in this league might include the Hoover Dam, the Golden Gate Bridge, and massive buildings, such as the Khobar Towers in Saudi Arabia and the World Trade Center in the United States—both targets of terrorist attacks. What potentially qualifies these as monumental technologies is the fact that their dramatic impacts are more proportional to their costs than are any quantified benefits.

To be sure, the space program has had world-transforming practical benefits, notably in weather and communication satellites. In the 21st century, it is difficult to imagine a world without accurate tracking of hurricanes, instantaneous global communication, or information-at-my-fingertips through the World Wide Web. Whether the benefits of these justify the hundreds of billions of dollars spent on NASA over the past 50 years is a calculation that has never been attempted. The more resonant justification for the space program and similar explorations or salients is not that of cost/benefit calculations but of the vision of mankind's continual quest of the Endless Frontier.

Every innovation begins in a very tentative manner, in which some dreamer tries some new idea or device with a "Will this work?" attitude. If the visionary had a good, factual basis for knowing that the innovation would work, then it would not be an innovation. Studies may be commissioned, surveys conducted, models constructed, but ultimately there is a leap of faith involved in any innovation. Innovation is a negotiation with the future, or more accurately an attempt to redefine the future from contingency to control. Far from fatalistically accepting a future of "What will be, will be," innovation is today's attempt at controlling tomorrow. As with any negotiation, the outcome is always uncertain. Innovation is thus one more phase in the expansion of a capitalist economy. Having bestrode the narrow world, and delved into every corner of domestic life, the capitalist economy now reaches into the future.

AN ENDLESS WILDERNESS

If we accept that technology includes an extension of state and corporate power, then what happens to technology on its frontiers? What happens at the end of the dirt road, on the fringes where state power, a broad and mighty current at the capital, is reduced to an insignificant trickle at a some distance from the center? The modern state, with its conception of sovereignty within defined frontiers, is a recent invention, an alternative to dynastic realms and feudal warlords. These earlier regimes were characterized by a well-ordered center and a lawless periphery. Yet, even today, state power wanes in proportion to distance from the metropole. This is easily observable in such regions as rebellious Chiapas, in breakaway eastern Anatolia, in branch plants thousands of miles from corporate headquarters, and in the deserts of Arizona, where polygamous Mormon sects defy both church and state. Peripheral regions such as these, as Immanuel Wallerstein (1974) first described, are part of the modern world system.

What happens to technology on the fringe of corporate and state authority? To the extent that technology is a social product, created in the laboratories and workshops of core institutions and corporations, represented by authoritative standards, and diffused through social, industrial, and artifactual networks, what transformations does it undergo the further it gets from this core? This is the question of technological peripheralization, an important and intensifying issue as the world becomes more dependent on technological devices for its survival and more knit together by technological networks.

Just as the benefits of technology are distributed unevenly so, too, are its artifacts. Such matters as spare parts, skilled personnel, and user manuals, abundantly available in the core, are often in short supply on the periphery. Systems such as personal computers, which perform miracles in core regions, are often experienced as drags on productivity in regions where computer skills, computer parts, or reliable networks are in short supply. The uneven distribution of technology and its benefits means that on a global scale technological development creates greater complexity. This complexity, further, can be a source of instability.

If end-user devices (the artifacts that people actually use) are modular and systems are loosely coupled, then technological artifacts end up being used in surprising and creative ways on the periphery. For example, cellular telephones have diffused rapidly, once the transmission infrastructure was in place, largely due to the fact that the devices themselves were small and cheap, and offered service superior to the landlines (Miller and Horst 2006, for example). Third World countries such as Nigeria, which for decades had inferior landline connectivity, have "leapfrogged" over core countries, now having cellular service superior to that of the United States. With such flexibility, devices such as these find uses never imagined by their inventors: detonating booby traps, for example. Fishermen in Brazil, while still at sea, use cell phones to check on prices being offered for their catches. Farmers use laptop computers to track commodity prices on the Chicago Mercantile Exchange. The "democratization" of technology, which is supposedly a characteristic of the 21st century, is actually a democratization of end-user devices, inasmuch as the development and operational infrastructures are still firmly in corporate hands.

With large-scale, tightly coupled systems, the eclipse of state power on the fringe by local concerns means that advanced technological systems, particularly those that package large amounts of energy, information, or toxicity, are inherently dangerous. This is illustrated and confirmed by the safety record of civil aviation in Africa and Latin America. A system—commercial air transport—that is remarkably safe in core regions such as Europe and North America, is significantly more hazardous in peripheral countries. An accident rate of 0.39 hull-loss accidents per million departures in North America in 2006 compares with a rate of 9.65 per million in Africa: more than 20 times higher. This is not because the aircraft are inherently more dangerous in Africa, but rather because the diffusion of the service—air transport—outpaces the diffusion of the infrastructure of training, resources, facilities, regulation, and attitudes that make the service ultra-safe in the core. Matters as banal as runway fencing, taken for granted in the core, are at times neglected in countries outside the core, resulting in livestock collisions being a flight hazard unique to the Third World.

More generally, industrial safety (or its lack) in the Third World is one of the overlooked consequences of technological diffusion from the core to the periphery. Since artifacts (such as factories) and techniques (such as coal mining) are diffused more easily than infrastructure (such as good roads) or superstructure (such as public regulation), industries that in the developed world have achieved acceptable levels of safety remain hazardous in the developing world. Again, this is not due to any deficiency in the cultures or governments of the developing world, but simply due to the mismatch between the pace of invention (growing geometrically) and diffusion (spreading arithmetically).

The Endless Frontier that Vannevar Bush imagined as a bracing challenge to America's "pioneer spirit," is thus quite accurately an unbounded wilderness of insecurity, instability, and hardship. In contrast to the mythic frontier that was created only in Vannevar Bush's 20th-century lifetime, the actual frontier of the 19th century was an unstable state, characterized by hunger, disease, and hostility. Building a stable community was a generations-long affair, which on the edges is still not accomplished. From this perspective, the metaphor of "Endless Frontier" takes on a new coloration. For Vannevar Bush, the Endless Frontier was the open-ended character of human knowledge, translated like the practical man he was into an open-ended character of human mastery—whether over disease, the forces of nature, or the planets in the heavens. Missing, in Bush's vision, was control over the wildness within, the habits of the heart, the passions that can inflame people individually or collectively to do great or terrible things. Patronizingly, Bush (1945) excluded the social sciences from his vision of the Endless Frontier. "Progress in other fields, such as the social sciences and the humanities, is likewise important; but the program for science presented in my report warrants immediate attention" (letter of transmittal). In July 1945, this statement was supported less by wartime exigencies and more by institutional ambitions. Yet, as we shall see in the next chapter, the failure of culture, broadly construed, to keep pace with the increasing capabilities of mechanical, electronic, and nuclear devices has created the far more fragile technological world in which we live today.

Note

[1] My italics and exclamation point is an imitation of the Futurists, who had the idea that "of course!" every innovation would be instantly adopted.

[2] The Congressional Office of Technology Assessment was established in 1972 to provide the Congress with expert, nonpartisan analysis of technological possibilities. Its studies of acid rain and health care are widely regarded as setting a high standard for evaluation of technology. When a Republican insurgency took over the Congress in 1994, the OTA was abolished.

Chapter Four

Technology Stumbles

> We nuclear people have made a Faustian bargain with society. On the one hand, we offer . . . an inexhaustible source of energy. . . . But the price that we demand of society for this magical energy source is both a vigilance and a longevity of our social institutions that we are quite unaccustomed to. (Weinberg 1945[1994:front flap])

At the conclusion of the Second World War, where Vannevar Bush saw an "Endless Frontier," Alvin Weinberg, the Director of the Oak Ridge National Laboratory where nuclear reactors were developed, saw a pact with a malevolent Power, the bill for which would eventually come due. The demands this bargain would place on social knowledge, insight, and institutions, dimly understood at the time, required another 30 years of global industrialization and cultural globalization to come sharply into focus.

Through the 1950s and 1960s, the electric power industry embarked on an aggressive program in nuclear power plant construction. The Atoms for Peace program, proclaimed by President Dwight Eisenhower in 1953 in a speech before the United Nations to a world in which the horrors of Hiroshima and Nagasaki were still fresh memories, provided an optimistic hope that the destructive potential of nuclear energy, and human ingenuity in general, would "not be dedicated to his death, but consecrated to his life." The chairman of the Atomic Energy Commission predicted that future generations would "enjoy in their homes electrical energy too cheap to meter" (quoted in Laurence 1995:1).

On March 28, 1979, a reactor operated by Commonwealth Edison of Pennsylvania experienced a near-meltdown. A valve malfunctioned, draining water from the core, causing it to overheat. The operators were confronted with confusing readings from multiple instruments, and they shut off the emergency pumps. This caused the coolant level to fall and the temperature to rise, approaching the 5,200-degree level that would have caused the reactor to melt through the floor of the containment building. For 16 hours confusion reigned until the designer of the reactor, the Babcock & Wilcox Company, got word through to the control room to turn on the pumps, at which point the temperature began to fall.

If the operators in the control room were confused, the managers, regulators, and public authorities were even more so, a fact that turned this serious incident into a major disaster. Plant managers denied that there was any danger, even as they advised pregnant women and children to evacuate. The Nuclear Regulatory Commission required hours to relay emergency messages. The mayor of Scranton expressed his frustration that he could not get straight answers from the utility company that operated the plant. What the public saw was a collapse of government, both public and private. This loss of confidence and trust that the utility and the public authorities knew what they were doing with a very powerful technology was the true disaster of Three Mile Island.

In this chapter, I examine the failures of technology: airplane crashes, chemical plant explosions, structural collapses, and software failures. Technology, of course, refers not only to electromechanical and cybernetic devices but also to the operators who control them, the executives who guide their enterprises, the public officials who regulate them, the institutions that embrace them, and the cultural values, however contested, that ultimately guide their use. Technology's failures, as I document here, are no less instructive of the cultural significance of technology than its dazzling successes.

THE LIMITS OF MATERIALS, HUMANS, AND INSTITUTIONS

From a broader, historical perspective, the near-meltdown at Three Mile Island is a perfect example of the unexpected challenges that arise when technologists test the limits—"push the envelope"—of known operating regimes, limits that are discovered only by stumbling over them. The operating regime includes both the physical properties and environment of the technology and the social and institutional spaces that it occupies.

Failures of materials, involving physics and mechanics, are easy to understand and are described in this section. In the first era of commercial air travel, the upper limit on speed was 200 mph—the cruising speed of the DC-3, whose maximum altitude was 23,000 feet. In the same decade that the DC-3 first took flight, Frank Whittle and Hans Ohain separately invented the jet engine. By 1941, jet engines had been installed on experimental aircraft and were first used in an operational system in 1944: the German V-1 "Buzz bomb" rocket (actually an unmanned aerial vehicle) that terrorized London in the closing days of World War II. After the war, the aircraft manufacturer de Havilland developed a commercial airliner, the Comet, that went into service in 1952. The Comet flew faster and in thinner air than previous airliners. This was a flight regime that was unfamiliar to its designers and that placed new stresses on its fuselage. At the corners of its square window panels cracks developed, which on several flights led to the breakup of the airplane in flight. After several well-publicized catastrophic accidents, costing hundreds of lives, the Comet was taken out of service and redesigned; later generations of jet airliners learned from the Comet's flaws.

Limits can also include limits on human capabilities and comprehension. In the 1960s and 1970s, aircraft manufacturers began developing "fly-by-wire" aircraft, which would use electrical signals instead of the cables and levers of hydraulic controls that were standard on all airliners since World War II. "Flight management systems"—sophisticated computers that managed the controls and navigation of the airplane made flying an airliner less like driving a car and more a matter of inputting data into a computer. Paradoxically, these systems place *greater* demands on the flight crews for coordination, communication, and situational awareness, creating a space where small errors could have large consequences. This fact was made clear on January 20, 1992, when an Airbus A320, one of the earliest airliners to integrate flight management systems with all aspects of flight, crashed into Mt. Ste Odile near Strasbourg, France. A subsequent investigation found that the most probable cause of the accident was the pilot's inputting a "vertical descent speed" of 3.3 (that is, 3.3 thousand, or 3,300 feet per minute) when he should have been inputting a *glide slope* of 3.3 (that is, 3.3 degree angle of descent, which would correspond to a vertical speed of 800 feet per second), thus causing the aircraft to plummet abruptly. This momentary confusion over a digital input might be familiar to anyone who has fumbled with a multi-featured cell phone, or attempted to program a VCR. Ambiguous communication with the control tower, a relaxation in situational awareness, and a sudden increase in workload, also contributed to an accident that cost 87 lives.

As technological devices become more complex, they make greater demands on their operators, so that a standard part of the training of commercial flight crews is in the interpersonal skills of teamwork, or "Crew Resource Management." These skills include situational awareness, assertiveness, task sharing, workload monitoring, and other *interpersonal* skills. Several accidents in the 1990s and the first decade of the 21st century, all involving technologically sophisticated airliners, were traced to a breakdown in coordination either among the flight crew or between the flight crew and air traffic control.

Technologies can also push the envelope of institutional capabilities, although these are more controversial, inasmuch as interested parties have financial and emotional investments in existing institutional arrangements and will vociferously defend their investments. In an individualistic culture where elected officials routinely deny even the concept of public responsibility, identifying technology-induced institutional failures is especially difficult, and blaming the operator of the system—"human error"—is a frequent response to mishaps. Institutional controversies over technological threats include the disposal of radioactive wastes, damage done to the environment by pesticides, alterations of the food chain with genetically modified organisms, and damage done to the planet's climate by carbon dioxide emissions. The very fact that each of these is an ongoing controversy demonstrates that our educational and political institutions are slow in reconciling conflicting social values (social goods vs. private profits, efficiency vs. safety, or security vs. civil liberties, for example) when technological developments disturb a previous equilibrium.

For unambiguous cases of institutional failure in the face of technological threats we might look at some rudimentary, man-made structures whose failure reveals interplays of institutions and artifacts, unleashing the destructive power of millions of gallons of water. Dams, levees, and similar impoundments rarely come to mind when one thinks of technology, even though they share all of the features that we have learned to associate with technology: they are human-built, usually according to engineering standards, and they harness enormous amounts of energy. The Hoover Dam, for example, whose construction was completed in 1935, impounding Lake Mead with more than 9 trillion gallons of water, produces in a typical day power on the same order of magnitude as the Hiroshima nuclear explosion. When insufficiently engineered, or insufficiently regulated, concentrations of energy of this magnitude have enormous destructive potential.

The nation had an early lesson in this destructive potential on the night of May 31, 1889, when a dam on the South Fork of the Conemaugh River failed, sending a 40-foot wall of water rushing down the valley, obliterating the town of Johnstown, Pennsylvania, 15 miles below, killing more than 2,000. The dam was a hastily built earthen dam, used to construct a summer resort for the Carnegies, Fricks, Mellons, and other tycoons of the Pennsylvania coal country. In the words of a reporter who investigated the disaster,

> There was no massive masonry, nor any tremendous exhibition of engineering skill in designing the structure or putting it up. There was no masonry at all in fact, nor any engineering worthy of its name. The dam was simply a gigantic heap of earth dumped across the course of a mountain stream between two low hills. (McCullough 1968:245)

Although the Johnstown Flood was called a "natural disaster" by some, the fundamental cause was the failure of a human-built structure. Reviewing the damage, John Wesley Powell, the founder of the U.S. Geological Survey and a sometime ethnologist concluded, "Modern industries are handling the forces of nature on a stupendous scale. . . . Woe to the people who trust these powers to the hands of fools" (Powell 1889 in McCullough 1968:263).

Culture is about learning. It is also about learning impairments, since every culture creates blind spots among its members. Two of the blind spots in American culture are that powerful, socially respected persons and groups might actually do great damage to society and that passive structures can pose active threats. The Johnstown Flood was re-created 83 years later on February 26, 1972, when a slag dam in West Virginia failed in a heavy rainstorm, sending a 30-foot wall of water down Buffalo Creek, killing 125, destroying a thousand homes, and leaving more than 4,000 homeless (Stern 1976). In this dam failure, conditions practically identical to Johnstown prevailed: a poorly designed structure, with almost no engineering calculations of the stresses created by large impoundments of water, no construction of floodgates to relieve excess pressure, and the stripping of the hillsides above the dam, leading to excessive runoff in a rainstorm. The Pittston Coal Company, which was responsible for the dam, was sued in several actions for nearly $400 million in damages but ended up paying approximately $19 million.

In the last century, the art of social responsibility failed to keep pace with the art of structural engineering. On August 29, 2005, a levee failed along the Mississippi River. It had been built by the U.S. Army Corps of Engineers, one of the most sophisticated construction engineering organizations in the world. From a technological point of view, the loss of more than 1,800 lives and the devastation of New Orleans in the course of Hurricane Katrina are less tragic than melancholy, inasmuch as one sees a repetition of the *institutional* failures that accompanied other failures of passive structures. These include a failure to assess and communicate impending hazards when extraordinary events, whether a thunderstorm or a hurricane, threatened the impoundment.

Numerous other structural failures around the world make clear that even "low tech" structures can be as dangerous, if not more so, than highly engineered devices, particularly if the "low tech" structures are presumed safe. On October 9, 1963, a rain-soaked *hillside* above the Vaiont Dam in the Italian Alps failed, destabilized by the hydrostatic pressure of water in the impoundment above the dam. The hillside's failure sent 312 million cubic yards of rock and debris into the reservoir, half-filling the reservoir and creating a wall of water that over-topped the dam, rushed down the valley, and killed approximately 3,000 people. Amazingly, the well-constructed dam held, even as the hillside failed (Hendron and Patton 1986).

Another failure of a "low-tech" structure was the 1981 collapse of a skywalk in the Kansas City, Missouri, Hyatt-Regency Hotel. The hotel had an innovative structure, an atrium, surrounded at multiple levels by skywalks. During a dance and concert in the atrium, the excessive weight of partygoers on the skywalk caused a collapse, killing 114. A subsequent investigation revealed that the design of the skywalk included an impossible-to-construct unified box-beam suspension, leading the builders to use a far weaker construction (Hauck 1983). One might also consider the Tacoma Narrows Bridge, completed in 1940, which combined a bridge deck more delicate than normal practice with a suspension across a large waterway. When a high wind came up shortly after the bridge was opened, wind-induced harmonic vibrations (which had *never* been a concern for bridge engineers) resulted in the bridge first swaying, and then rhythmically undulating in the wind, before it collapsed. Fortunately, no lives were lost, and the collapse supplied some memorable film footage of "Galloping Gertie," as the bridge was subsequently named.

What all of these had in common was that they were innovative structures that pushed the envelope of engineering knowledge and institutional capabilities, just as the Comet airliner pushed the envelope of material capabilities. New technologies push into the thinner atmosphere of regulatory voids, and public institutions frequently scramble to keep up with the technological developments only after a dramatic loss of life.

Analyzing institutional failures takes us out of the realm of technology and into the realm of culture, or more accurately, into the realm of political analysis, verging on a register of the crimes, follies, and misfortunes of mankind.[1] Are wars an example of technologically induced institutional failure? Many have considered the First World War in such terms (Kern 1983), noting that the telephone and telegraph relayed messages across Europe faster than governments could absorb them, and railways speeded up troop mobilizations, creating facts on the ground that statesmen spent precious days trying to comprehend. The introduction of a new military technology—aerial bombardment of civilians—turned the Spanish Civil war from a national squabble into a trial heat for World War II.

Technology carries humanity along faster, higher, and with greater force—the *citius, altius, fortius* of the Olympic motto—and thus ventures into new environments, whether the depths of the ocean, the thin air of the stratosphere, or a new social topology of rapid, far-flung, and unanticipated consequences. Further, it is doing so at an ever faster pace. In 1889, seven years before the first modern Olympics coined the phrase above, the historian Henry Adams, awestruck at the Machinery Hall at the Columbian Exposition in Chicago, sketched out an equation that the amount of energy harnessed by humans doubled

The Tacoma Bridge across the narrow strait in Tacoma, Washington, before and after a large section of the concrete roadway in the center span of the bridge hurtled 200 feet into Puget Sound on November 7, 1940. High winds caused the bridge to sway, undulate, and finally collapse under the strain.

roughly every 10 years (Adams 1961[1918]:490). He based this calculation on an observation of a 4000-watt dynamo at the exposition, tracing
back to the first locomotives earlier in the century. Two decades later,
the *Titanic*, with its two Olympic-class triple-expansion 16,000 horsepower steam engines, was on the leading edge of this curve of doubling
horsepower, as was the thousand-megawatt reactor at Three Mile
Island 65 years later. Adams, of course, anticipated by 70 years the discovery of another "law" (actually a descriptive generality, inasmuch as
it describes only the results but not the underlying dynamic) of technological acceleration, "Moore's Law." Gordon Moore stated in 1965 that
the capabilities of microprocessors doubled every 18 months. His pre-
dictions, contained in "Cramming More Components onto Integrated
Circuits," (Moore 1965), actually failed to anticipate the transformations in everyday life that computers would create over the following
decades. These undeniable facts of technological acceleration, themselves a product of the topologies of contemporary sociotechnical networks, create new stresses in society when families, industries, and
even entire communities attempt to keep up with them.

  In sum, the acceleration of technological development means that
technology can be expected to continue to challenge the physical and
cultural limits. "Appropriate technologies"—those devices designed for
the Third World where physical and institutional limitations are even
more demanding—may well have lessons for avoiding further catastro-
phes in the First World.

INDUSTRIAL FAILURE

  The accelerating speed, altitude, and force afforded by modern
technology accrue to those who control it, which increasingly means
large corporations. The continental and later global reach of these corporations are world transforming, a 19th-century vision of Goethe that
in the 20th century was named "globalization." This acceleration of
social force was discovered in the 19th century when new technologies,
the railroad and the telegraph, gave birth to the modern corporate
form. Prior to the rapid transportation and communication that these
afforded, businesses were almost entirely local affairs, commanded by
a proprietor assisted by a few clerks. Far-flung trading companies, such
as the Hudson Bay Company, might command a small army of agents,
but these agents were more of the nature of independent contractors
rather than company employees. An imperatively coordinated business
operating across multiple time zones (an invention that followed railways) yet unified by standard devices required a predictable administrative apparatus that the modern corporation, with its multiple

operating divisions, its administrative staff, and its functional departments, supplied.

Technology thus creates concentrations of power—not only expressed in megawatts but also in market shares, employment contracts, and public officials suborned—that the founders of the American Republic could scarcely have imagined. Occasionally these concentrations of power do great harm. On April 16, 1947, two ships, the *SS Grandcamp* and the *SS Mellencamp*, were loaded with ammonium nitrate in the harbor of Texas City, Texas. The cargo, a fertilizer intended for war-torn Europe, was known to have explosive potential, but the regulations for handling explosive cargos were nonexistent or weakly enforced. When a fire broke out in the hold of the *Grandcamp*, the captain's actions, spraying water into the hold, created a heat-producing reaction that soon triggered an explosion with a force similar to that of a nuclear bomb. The Texas City explosion, which leveled large parts of the city and cost at least 583 lives, remains the worst industrial disaster in American history. Even in peacetime, death and destruction can now be delivered on a massive scale.

In terms of loss of life, the worst industrial disaster in world history took place on December 13, 1984, in the Indian city of Bhopal. A pesticide plant, located in India and near a shantytown as cost-cutting measures, had an explosion, causing highly toxic methyl isocyanate (MIC) to leak into the air. Prevailing winds carried the toxic cloud over the shantytown, eventually killing more than 5,000 villagers. A subsequent investigation revealed a long list of malfeasances, malfunctions, human errors, and institutional failures: the plant had been shut down for cleaning, but it still contained more than a thousand gallons of MIC. Failure to insert a "slip bind" led to the MIC leaking into a water tank, where it reacted, causing the explosion. Far more telling than these human errors were the institutional failures. The safety manuals were in English, a language that many of the personnel did not read. The operators' morale was low, in part because their numbers had been cut back and in part because they were occasionally penalized for following procedures; unsurprisingly, when the plant went out of control, the operators left the scene. A top-level discontinuity in management had the consequence of nonenforcement of safety practices and safety culture. The low priority of the plant for Union Carbide also explains why these conditions persisted in a dangerous operation (Shrivastava 1987).

A common feature of Texas City, Bhopal, Three Mile Island, and many other industrial accidents is that they are "normal," i.e., expectable, in the words of Yale Sociologist Charles Perrow (1984). A "normal accident," in Perrow's view, is one that should have been expected, given the design of the technological system. Perrow focuses on two aspects of design that are found in many sophisticated systems: complexity and tight coupling. Complexity is the difference between an Airbus A320, with its computerized systems, and the unsophisticated DC-3, perhaps the most robust, widely flown aircraft the world has known. Complexity is measured in the proliferation of states, procedures, subsystems, and user inputs in advanced systems. The more complex a system, the greater the number of failure modes it has, a fact that system designers understand and work to minimize.

"Tight coupling" refers to the interaction of these states, procedures, subsystems, and user inputs. From one perspective, tight coupling is elegant engineering or efficient operation, cramming more passengers into a cabin or more airliners into the skies around a terminal or more bags of explosive into the cargo hold of a ship. From another perspective, tight coupling is an invitation to a disaster, since the deviation or failure of one component invariably cascades into other fail-

The nose of a Continental Airlines DC-10 reflected in the pool of rainwater on the runway at Los Angeles International Airport, after it crashed on takeoff. Passenger airliners are complex and tightly coupled yet ultra-safe systems. The attention given to their occasional crashes is part of the social definition of technology in today's society.

ures. It was tight coupling—two 747s diverted to a small airport in Tenerife—where an overcrowded terminal and an overcrowded schedule created pressures for a premature departure that caused a runway collision of two aircraft in March 27, 1977, killing 583. Tight coupling can also be as ordinary as the small buttons on a cell phone, leading to misdialed calls, or as commonplace as a traffic jam during rush hour.

When tight coupling is joined to complexity, accidents are expectable. When large amounts of energy or toxicity are included, the accidents can be catastrophic. Institutions and industries learn this one by one and adopt measures to minimize accidents—assure high reliability—in the face of great hazardous potential. These measures, which I examine more closely in chapter 5, include frequent training, devolution of operational authority, flexible communication, and adequate staffing. They also include strict, by-the-books regulation. All of these measures were stinted at Bhopal. As was the case with Bhopal, competitive pressures frequently force industries and governments to cut back on training, cut back on staffing, cut back on supervision, and cut back on recovery resources, rolling the dice, hoping that the inevitable mishaps do not escalate into major disasters.

It is a truism that technology has made the world "smaller"—that is, more tightly coupled. Technology has also made the world more complex, cramming more specializations, more lifestyles, more agendas, and more divergent loyalties onto the small surfaces of corporations and communities. Occasionally, this organizational complexity leads to a disaster, as when miscommunication between contractors led to the explosion of the Space Shuttle *Challenger* (Vaughan 1996).[2] More typically, it means that only the most fortunate corporations and communities will be able to keep up with the wild ride of technological acceleration.

INFORMATION TECHNOLOGY
SUCCESSES AND FAILURES

As technology harnesses great physical power and cybernetic force, it creates expectations that at times outpace what technology can effectively deliver, or innovations that outpace what institutions can absorb. This race between promise and performance was played out in the massive investments in factory and office automation of the 1990s, investments that have not abated in the current century. In the 1980s and 1990s, office work was revolutionized as desktop computers, networks, and increasingly powerful applications gave office workers new capabilities and new responsibilities.

The effective use of technologies such as office computers and touch-screen devices comes not from their ability to speed up existing

processes—producing scrap more efficiently, in a memorable phrase—
but from their ability to enable new processes and new business models, guided by new cultural understandings. It is a break with old business models that enabled Dell to drive down the price of personal
computers and Wal-Mart the price of everything else, and Microsoft to
get into the Internet age by setting a price point—free—for their Web
browser, Internet Explorer. In fact, it has been Microsoft's intuitive
understanding of the new software economics, more so than the quality
of its products, that enabled it to dominate the personal computer software market.

Throughout the 1980s and 1990s, the explosive growth of office
automation created the amusing spectacle of old-school executives
attempting to grasp a new-fangled gadget, the desktop computer. Desktop processing afforded many local efficiencies, such as engineering
drawings produced more rapidly, memos revised with only a few key-
strokes, financial projections produced almost instantaneously, view-
graphs (also called, variously, pneumographs, slides, transparencies,
overheads, viewfoils, or PowerPoint) produced by anyone. For every one
of these local efficiencies there were two steps backward in overall
effectiveness: different teams of engineers insisting on reworking the
drawings handed to them, middle managers lovingly doctoring their
memos with artwork borders and 𝖿𝖺𝗇𝖼𝗒 𝖿𝗈𝗇𝗍𝗌, a profusion of neatly
presented financials based on sloppy assumptions, and hours con-
sumed in boring PowerPoint presentations.

All of this computational churning added up to the "productivity
paradox" that the economist Robert Solow first noted in 1987: "You see
the computer age everywhere but in the productivity statistics" (Solow
1987 quoted in Triplett 1999).[3] This famous observation, which has
been very productive of articles in economics journals, actually stems
from the transformational nature of computing. Small, cheap, ubiquitous computers do not simply speed up or make more efficient the control of the old economy; they create a New Economy, in which
information and attention to information have become commodities. As
Feenberg notes (1999:90), comparisons of efficiency and productivity
are valid only when they are made for similar systems ("within a paradigm," in Feenberg's terms), such as miles-per-gallon of automobiles.
Comparing miles-per-gallon of automobiles and freight trains makes
little sense, or transcription speeds of quill pens and PDAs, because
these two devices are used for completely different purposes. Likewise,
comparing information productivity in an economy where information
is used primarily for controlling material production, and information
productivity in an economy where both information and the viewing of
information have become commodities, has little to recommend it.

The productivity paradox was based on the factual observation
that measurable improvements in productivity were not commensurate

with measured investments in information technology, a mismatch usually suggested by measurement errors, time lags in the diffusion of innovations through institutions, or inefficient use of the technology as described in the examples above. For example, e-mail was developed in the 1970s, and the fundamental standards of the Internet were laid down in the 1980s, but it was not until the mid-1990s that businesses began making widespread use of both. The failures of office automation projects (Gartner Group 1999; Gillooly 1998) are also noted, as are entire classes of technologies, such as software applications for Enterprise Resource Planning, a business management strategy that integrates different departments and functions into a single component, whose implementation is notoriously difficult and whose botched installation has bankrupted entire companies (Robbins-Gioia 2002; Scott 2000; Standish Group 2001; Vidyarana and Brady 2005). These failures are not so much the failures of technologies or of users. They more often arise from the birth pangs of new ways of doing business, where images, attention, and even mouse-clicks have become commodities to be bought and sold; where network externalities (the benefits that accrue by linking up with a network) and increasing returns overpower the diminishing returns of traditional economic expectations (Arthur 1994).

At times, technologies collide with institutions and with the deeply ingrained assumptions institutions embrace. This was illustrated by the failed effort of a strong but rigid institution, the Federal Bureau of Investigation, to enter the computer age. In the wake of September 11, 2001, and the documented failures in information sharing that allowed the terrorists to evade detection, the FBI undertook a $170 million modernization of its information systems. The Trilogy project consisted of the acquisition of new computers, the upgrade of the FBI's network, and the installation of a new system, the "Virtual Case File," that would allow FBI agents to share case information both among themselves and with other agencies. Three years later, after spending more than $104 million, the FBI abandoned the project. Numerous congressional, journalistic, and scientific inquiries (GAO 2006; McGroddy and Lin 2005) pointed to a litany of shortcomings, not the least of which was that, culturally, the FBI was still in the typewriter age. As recently as 2005 the FBI's case records were paper-based, with a mainframe computer (accessed through outdated terminals) that served primarily as a storage device for the paper records (Knorr 2005). Whether the FBI as a law-enforcement institution still living in the house of J. Edgar Hoover can adapt to the new millennium of networked terrorism is a troubling question.

Another technology stumble of the same decade was touch-screen voting. America had been voting with paper ballots for centuries, the counting of which is a laborious and time-consuming process. Mechanical voting machines were prone to breakdown, and local officials in 50 states and thousands of municipalities all had their own preferred solutions in optical scanning or punch cards. When touch-screen voting machines were first produced, they seemed an ideal solution: the near-ubiquity of automatic teller machines suggested that voters would readily take to these new devices.

They didn't. The laws and data structures that made ATMs robust—paper trails, accounts to be reconciled, and a requirement that the bank justify its position in the case of a dispute—were not found in the environment of contested elections, secret ballots, and volatile memory chips. Touch-screen voting machines were introduced at a time when levels of mistrust were rising in American society, and the sacredness of ballot integrity meant that these devices would never be justifiable. The controversies over touch-screen voting in the 2000–2006 elections prove the limits of technology when confronted with the stubborn facts of social division.

Investments in technology—whether investments of personal time or public or corporate resources—come at the expense of investments in social capabilities. Every minute spent learning how to read the face of an instrument is one minute less spent learning how to read the face of another person. Yet, when technology creates social complexity and multiplies divisions within a society, cramming more loyalties and lifestyles into organizations and communities, then learning how to read others' faces, learning how to interpret a tone of voice or respond to a gesture of another's hand, is just as important a skill as operating a computer or navigating the World Wide Web. As airline pilots have learned with Crew Resource Management, sophisticated technologies require sophisticated social skills.

It is this lack of interest in learning others' faces, substituting technological wizardry for civic self-awareness, diplomatic skill, and honest communication, that lies behind another techno-stumble, the "Virtual Fence." Part of the "Secure Border Initiative," the Virtual Fence is a two-hundred-mile stretch of sensors, communication towers, physical obstacles, unmanned vehicles along the Mexican border, meant to secure a country with 6,000 miles of borders and more than 10,000 miles of coastline and uncounted millions of undocumented immigrants who have entered the country legally. The Virtual Fence is a perfect example of what Alvin Weinberg called a "techno-fix"—the application of a technological solution to a problem that is inherently nontechnological. Techno-fixes are sufficiently meager in their accomplishments and expensive in their costs—$1.5 billion for the Secure Border Initiative— to permit one to wonder what sort of problems they actually fix. The massive investments of public resources and political capital in the Virtual Fence assure that its failures will be dealt with in the manner of many other prosaic organizational failures: trumpeting small-scale achievements, downgrading the objectives, or changing the subject.

THE TRAGEDY OF THE COMMONS

In the 1960s, the American public began to worry about the effects of technology and the limits of technological possibilities. Rachel Carson's *Silent Spring*, published in 1962, warned readers of the dangers of DDT and other pesticides, eventually leading to the banning of DDT and tighter restrictions on other pesticides. Paul Ehrlich's 1968 book, *The Population Bomb* warned of the dangers of runaway population growth. Concern over nuclear weapons led Jerome Wiesner and Herbert York in 1964 to write that the dilemma of increasing nuclear power and decreasing national security *"has no technical solution. If the great powers continue to look for solutions in the area of science and technology only, the result will be to worsen the situation."* A few years later, Garrett Hardin, in "The Tragedy of the Commons," citing Wiesner and York, argued that "The population problem has no technical solution; it requires a fundamental extension in morality" (Wiesner and York 1964; Harden 1968). This challenge of technology, not simply to industrial performance but also to the social order, is considered below.

Hardin's perspective, "the tragedy of the commons," captures the limits of human rationality when people are required to share a finite resource. A commons—whether a meadow in which a village's sheep may freely graze or an atmosphere from which everyone breathes—is a limited resource that can be destroyed by overuse. Up to a point, natural resources are self-regenerating: grass in the meadows regrows every season, and photosynthesis restores oxygen to the atmosphere. When a resource is stressed beyond its self-regenerating capabilities, as is happening with grasslands in sub-Saharan Africa and with many fisheries around the world, the cycle of regeneration collapses. Yet, neoclassical economics argues that rational individual behavior consists in maximizing one's own benefits from the resource; in the neoclassical economist's view, people are limited only by the irrationality of morality. As technology has pushed back the limits of human activity, this paradox has puzzled orthodox economists.

Anthropology, which grew up studying communities that had learned to live within their limits, supplies an answer to the economists' puzzlement in the concept of culture. More specifically, by seeing how a shared morality of mutual obligation provides cohesion to a community we can understand the "irrationality" of living within limits. Although we should not romanticize so-called "primitive" (i.e., indigenous) communities, some of which in years past did exhaust their resources (Diamond 2005), most small-scale societies demonstrably did maintain equilibrium with their natural environment and with their neighbors, even though that equilibrium might include engaging in occasional warfare and raiding.

Within human communities, *trust* is a commons. Trust is based on a shared topology of stable relationships and identities. We automatically trust those with whom we *feel* some shared kinship, even when we meet them for the first time. This shared kinship can be based on ethnicity, nationality, profession, education, employment, or a number of other totemic characteristics. Societies are made up of complex webs of trust and mistrust among groups, individuals, and institutions, learned over years of interaction. The learning accumulated in these networks, social capital, is a resource available to all—a commons. Within stable communities there are institutional resources, whether village elders or public officials, who respond when this commons is abused. The response—identifying the abusers and making them pay or suffer for their misdeeds—is a public ritual. As Durkheim noted, punishment is important less as negative psychological reinforcement to those who might do harm, and more as positive social reinforcement of the community's commons of trust. Maintenance and, when necessary, restoration of this social fabric are the obligations of the institutions of society.

Trust builds on histories of reciprocities. We learn to trust, not just that another will do us no harm but also the *identity* of those who will do us no harm: how they fit into our cultural schema of kinship, our consociates whose behavior is guided by the same norms as ours. The familiar anchors of identity—ethnicity, gender, social class—are also familiar anchors of trust.

Technological innovation changes the terms of this equilibrium. Social changes driven by technology reconfigure the stable institutional arrangements both democratic and despotic, thus requiring new learning to figure out who and what can be trusted. As technological change accelerates, the shocks to these social networks also accelerate, raising the question of the long-term resilience of a society experiencing repeated techno-shocks.

Bridges can be rebuilt in a few months. Unsafe cars can be redesigned better in next year's model. Airliners can crash, an investigation reveals the cause of the crash, corrective actions are taken, and life goes on. But the betrayal of trust requires years, lifetimes, or sometimes centuries to repair. As accelerating technological innovation erodes the foundations of trust within a society even as it expands the networks of social circulation, it raises the melancholy possibility of creating a cultural wasteland in which an atomized, fragmented society contemplates its world-transforming achievements.

When Alvin Weinberg spoke of nuclear technology's Faustian bargain, he had in mind its cataclysmic, physically destructive potential, the management of which required greater wisdom than humanity perhaps possessed. He did not foresee, more generally, the manner in which powerful energy, transportation, chemical, and information technologies might corrode the shared understandings, shared commons, and shared trust upon which society is built and governed. The challenges of technology, we can now see are more multidimensional and potentially more far-reaching than simply those of controlling nuclear reactions. They include the challenges of achieving insight at an accelerating pace, of harnessing vast concentrations of institutional power, and of restoring a commons of trust that we have been depleting.

"Things are in the saddle, and ride mankind," Ralph W. Emerson wrote in 1847, at the *beginning* of what we now understand as the technological era. A hundred years later, Daniel Bell described the problem of "Adjusting Men to Machines" as one of adapting humanity to the "basic institution of industrial society, the factory" (Bell 1947:79), suggesting that if there was any contest between humanity and technology, the contest was over, and technology had won. Reclaiming the cultural resources to tame the technology that rides mankind is the challenge to which anthropology must now turn.

Notes

[1] This wording is taken from Gibbon (n.d.:69). "Antonius diffused order and tranquillity over the greatest part of the earth. His reign is marked by the rare advantage of furnishing very few materials for history; which is, indeed, little more than the register of crimes, follies, and misfortunes of mankind."

[2] In 1986, the space shuttle *Challenger* exploded in midflight, creating a major setback for NASA. A subsequent inquiry revealed that the contractor for the solid rocket booster, Morton Thiokol, interpreted ambiguous communications from NASA as pressure to launch despite cold weather. Morton Thiokol gave the go-ahead to launch, and two minutes into flight a rubber O-ring, brittle from the cold, failed, causing the solid rocket booster to explode. Diane Vaughan, in *The Challenger Launch Decision*, used these events as a case study of organizational failure when confronted with novel and complex technology.

[3] This quotation seems to have achieved within academic literature the status of an urban legend, something that everyone knows exists, because their best friend's brother-in-law heard about it from his cousin's colleague's uncle. Various authors have cited the *New York Times Book Review* and the *New York Review of Books* as sources (Brynjolfsson and Hitt 2003; David 2000; Dutton et al. 2005; Gottinger 2003; Goyder 2005; Kling 1999; Nibourg 2005; Triplett 1999). My efforts to track it down in these publications have not been successful. I can authoritatively state that it was not published in the *New York Review of Books* on July 12, 1987, because the only number of the NYR published in July 1987 was on July 16 (v34, #12). Nor was it published in the *New York Times Book Review* on July 12, 1987, according to a search on the NYT Web site. I am contacting Professor Solow to see if he can provide me with an authoritative citation.

Chapter Five

The Human Factor

Cybernetics, the science of systems, informs us that within any complex system, the element with the fewest degrees of freedom exerts the greatest control over the entire system. Wherever men and machinery come together, whether in an industrial factory or a transportation network or a computer workstation, the relative inflexibility of the machinery compels the more flexible element, humanity, to adapt.

Rapid technological developments in the 20th century created the necessity for a science of adjusting men to machines, in Daniel Bell's phrase (1947). This was first discovered in the wartime mobilization of aviation in World War II, when more pilots' deaths were caused by errors in piloting than by enemy hostilities. Similarly, as computers proliferated into homes and workplaces and everyday activities, psychologists and computer scientists and others began to understand that productivity was lagging, not because of defects in the microchips, but instead due to mismatches in *human–computer interaction.*

In any problem of adjustment, there are two complementary strategies of adaptation: selection and modification. Since any group of either human users or machines contains variation, one can cull those that fit best and discard the rest. Landfills and junkyards are filled with the industrial scrappage of such evolution, although the human scrappage of a technological age should trouble us far more than it does. Alternatively, one can attempt to modify either the machine or the human in order to make the two better fit together. This dual problem, of designing a perfect machine to fit with imperfect humans, or selecting and modifying imperfect humans in order to fit with perfect technological devices, is at the heart of the science of human factors.

In this chapter, I examine the two sides of this adjustment, the selection and modification of human users and machines in order to get along with each other. I examine "human factors" most broadly, going well beyond its traditional domains of ergonomics and human–computer interaction (HCI), to consider in what manner humanity is capable of managing the devices created by modern industrial technology. After examining some quotidian human factors issues in physiology and psychology, I examine the domain in which the factor of humanity is most challenging: the emerging world of virtuality, artificial life, artificial intelligence, and computer-mediated reality. Drawing on sensory and cognitive abilities, and increasingly supplying bodily engagement, these systems have the ability to mimic, modify, and even replace cultural experiences. As technologies both hard and soft occupy larger spaces in social life, the adjustment of culture—authentic or synthetic—to technological requirements becomes a human factors challenge, the hidden tensions of which I examine in the section later in the chapter: "The Secret Life of the Human Factor."

SWITCHOLOGY

Whenever one steps up to a technological environment, whether the driver's seat of an automobile, the cockpit of an airplane, the control panel of a complex machine or factory, or even the keyboard of a handheld personal digital assistant (PDA), one is entering an environment for which biological evolution has left us unprepared. The effective use of these devices makes far greater demands on the perceptual, cognitive, and coordination capabilities of people today than those made on Ice Age hunters. The greater the technological sophistication of these devices, whether in terms of miniaturization, complex functionality, or aggregation of information, the greater the demands they place on the individual operator—the pilot in a cockpit, the office worker at a computer, or a teenager with a cell phone/camera/PDA/music-sharing device at her ear.

As a scientific discipline, human factors emerged out of aviation, which uniquely placed hundreds of thousands of aircrew in uniquely demanding environments with uniquely demanding tasks. The challenge of controlling an object in three axes of motion at several hundred miles per hour, while breathing thin air in cold temperatures, placed impossible strains on the thousands of fliers who died before 1940, usually due to their own errors in judgment, perception, or control. A task as rudimentary as knowing which direction is up, normally unremarkable, is impossible inside a sky full of thick haze.

Out of challenges such as these emerged the discipline of human factors. Some of the earliest human factors work consisted of supplying pilots with sources of flight information through instrumentation, protection from the elements through prostheses (helmets, goggles, oxygen masks), and sources of revenue through passenger accommodations. This balancing of human capabilities, information, procedures, and the evolving technology in a challenging environment has typified human factors work ever since. As flight evolved from the province of daring young men and women to a budding industry, selection, training, and evaluation of pilots was a major task of human factors specialists. As the industry pushes the envelope of known performance limits, with lengthy flights, demanding schedules, stressed employees, packed cabins, smaller seats, and surly passengers, studies of fatigue, performance, and morale under stress are major human factors issues.

Within aviation, there are two significant currents in human factors. The dominant current focuses on the individual, whether crew or passenger, and his or her adaptation. Once the flying community had learned the basic physical limitations (do not fly above 14,000 feet without oxygen, for example), the focus of human factors became a matter of fine-tuning the physical and informational environment of the pilot for peak performance. How much information, for example, should be made available to the pilot? How should it be displayed? On an analog dial (with one or two pointers?), digital display, or a television-like screen? What should be the layout of instruments? What should be the shape of switches and control levers? What should be within reach of the pilot (and how long is the average pilot's arm)? How many audible alarms should a flight deck include? Questions such as these, which human factors specialist Ron Westrum has referred to as "switchology" (Westrum and Adamski 1999), are questions relevant not only to aircraft but also to the control of any sophisticated technology: nuclear power plants, locomotives, air traffic control towers, all have complex control panels and instrumentation. In the incident at Three Mile Island, for example, the fact that 18 alarms were sounding at the same time made it *more* difficult, not easier, for the operators to understand the problem. The psychological studies that addressed questions such as these were easily understood within a culture that emphasized the capabilities of individuals.

The individualistic approach to human factors is associated with a perspective that associates industrial mishaps with "human error," a term that is more accurately glossed as "operator error." Seventy percent of all aviation accidents are attributed to "human error." Design flaws, such as confusing instructions or complex control modes, are not classified as "human error," even though, presumably, designers are human. Nor are management errors, such as inadequate training, skimping on safety measures, or unrepaired equipment, considered "human error," even though managers are also human. The "human factor/error" perspective is a discourse that focuses blame on the least-powerful elements of complex technologies: those who operate and repair them. As we will see, these frontline workers have irreplaceable roles in assuring the safe and efficient operation of complex technologies.

A separate perspective examines the interactions between machines and human groups, whether the crews that operate them, the organizations that develop, manage, and maintain them, or the publics that use them. A growing body of literature since the 1980s has been examining safety and system performance in terms of *organizational factors,* including the cultures of organizations, rather than individual characteristics. Complex organizations and enterprises direct the design, operation, and maintenance of complex technologies including power grids, air transport networks, and industrial supply chains. These organizational complexes typically bring together numerous institutional, professional, and industrial complexes into a common field of operation. This *institutional superstructure* is populated both by human beings and by increasingly automated control systems. Any

Airbus A320 cockpit, an example of a highly automated airliner.

technology must be evaluated not only in terms of the actual user device (computer, automobile, airplane, locomotive) but also in terms of its institutional context. It is the institutional fragmentation of the health care industry in the United States, and not the shortcomings of medical personnel or the technology *per se*, that has hindered the adoption of electronic medical records in the United States.[1]

As computers become increasingly ubiquitous, finding ways to safely and intelligently use them is a management challenge. Modern corporations depend on large-scale networks intended to bring order to complex, sprawling enterprises. However, these systems create new forms of limited vision for executives. Spreadsheets and field reports available to executives can often be misleading. Old school, hands-on executives drive global enterprises into the ground, unable to read the spreadsheets and field reports that are the dials and gauges and warning lights of a global enterprise. Controls and instruments on modern airliners resemble video arcade games more than the knobs and levers of aircraft from the barnstorming era. Problems of human–computer interaction that are familiar to everyone, such as transposing digits on a keyboard, navigating through multiple screens, figuring out what the information truly *means*, whether by the executive in the boardroom or by the pilot at 35,000 feet, graduate from desktop inconveniences to matters of life and death.

LIVING ON THE SCREEN

Technology does not simply enhance our capabilities: it changes their character. The automobile not only extended the range of personal transport; it also changed the character of family and community connectedness in America. The computer, likewise, is not only accelerating the processing of information; its high-speed computation and global networks also create new possibilities of virtuality and artificiality as alternatives to reality, thus changing, perhaps irreversibly, the character of culture.

In their earliest days, computers were no more than information processors: fast, accurate, and stupid. In place of a company of clerks calculating payroll, computers supplied entire battalions of transistors, tabulating, recording, and generating paychecks, benefit deductions, mutual fund purchases, insurance premiums, federal, state and local taxes, and numerous other details that turn yesterday's pay packet into today's complex financial profile. Computers add velocity and control to these transactions. They also add complexity, although whether or not they add value is frequently debated. As long as "computer" equaled "mainframe,"[2] information processing was the extent of the computer's capabilities.

The character of any technology changes when it is democratized, when it moves from being the sophisticated tool of specialists into an appliance accessible to the masses. In the 1980s, the personal computer turned from a cool toy for geeks into a hot market for dozens of previously obscure companies: Osborne, Kaypro, Altair, Commodore, and many others supplied the PC revolution. In the 1990s, the Internet evolved from a rarefied ether where academic researchers convened into a digital commons whose population doubled every eight months.

Armed with personal computers and fused into a globally distributed information processor over the Internet with numerous human nodes, new communities and new life forms began to proliferate. One of the earliest combinations of bulletin board system (BBS) and chat room was the Whole Earth 'Lectronic Link (WELL), a San Francisco Bay Area virtual meeting place that grew out of the counterculture-inspired Whole Earth Catalog. The WELL enabled thousands of members scattered over hundreds of square miles to visit, exchange experiences, offer assistance in times of trouble, and coalesce into a "virtual community" that provided many of the functions provided by traditional, location-based communities (Rheingold 1993). Some of the difficulties in understanding these online collectivities as "communities" will be considered further in chapter 7.

The ubiquity of personal computers, at least in advanced industrialized nations, has enabled a proliferation of artificial or "virtual" realities: artificial intelligence, artificial life, virtual organizations, and the simulation of a broad range of capabilities formerly associated with "natural" organisms. Artificial Intelligence (AI), for example, reproduces within limited problem-sets capabilities formerly associated with human reasoning, including medical diagnoses, project management, energy exploration, mechanical design, and urban planning. Evolutionary systems explore, learn, and adapt to new problems such as optimal foraging within an ecosystem, or the stresses on structures such as bridges. Agent-based systems mimic and supply insights into the behavior of decentralized systems, whether flocks of birds (in the simulation "boids") or urban neighborhoods. One of these, Tierra, simulates the combination of randomness and teleology of evolution, by creating new (virtual) organisms in underdetermined responses to real-time environmental conditions. Virtual organizations design and manufacture products, coordinating the efforts of engineers, manufacturers, and distributors on every continent. If promotional accounts are to be credited, computers can do for humanity everything that culture formerly supplied.

As eloquently described by Sherry Turkle (1995), these systems test the boundaries of identity, self, community, and, by extension, humanity. Multi-User Domains (MUDs), for example, modeled after the campus game of Dungeons and Dragons, permit RL ("real life") users to adopt personae of various genders, ethnicities, and social capabilities, and thus "try on" alternative identities. In a MUD, anyone can be an Amazon warrior, a furry rodent, a witty raconteur, or an astronaut. Other interactive systems simulate psychoanalysis, marriage counseling, or self-help manuals. Programs such as the Sims™ replicate cities, markets, relationships, athletics, or running businesses. The intensity of emotional engagement that the RL users bring to these simulations and games creates a new perspective on the nature of the self and society, in which biological *Homo sapiens* might simply be one more component option in a cyborg matrix. ("Cyborg," a term coined in 1960, denotes a "cybernetic organism" that combines both animal tissue and computational devices. On cyborgs and anthropology, see Downey et al. 1995.)

Systems such as these have tested the boundaries of our understandings of "life" and (I suggest below) "culture." All of the characteristics associated with living biological organisms—learning, metabolism, reproduction, adaptation, evolution, defecation—have been simulated, thus suggesting that it is only a carbon-based naïveté that limits "life" to biological forms. The ability and the frequency with which users not only engage with these systems but also actually define themselves in terms of the systems, taking on roles from MUDs, likewise raises the question of where we locate the boundaries of "real."

Cultural infrastructures and institutional superstructures remain above and below all of these computational explorations, making it very unlikely that they can remake the human world. More precisely, most "virtual" or "artificial" systems work only in laboratory or institutional environments where some combination of adequate infrastructure, trained personnel, clear lines of authority, resource adequacy, and goal consensus can be taken for granted. Virtual organizations can design and manufacture products, because the goals are simple and clear, the personnel share a common industry culture and training, and the communication network supplies the gateway to membership. Virtual communities can provide meaning, belonging, and affect because their members share a larger set of cultural experiences, such as the 1960s counterculture in the case of the WELL. Thousands of artificial life forms, such as flocking boids or evolving Tierrans, can exist only under the sufferance of millions of RL designers in Palo Alto, programmers in Mumbai, assemblers in Guangzhou, warehouse workers at Wal-Mart, technicians in call centers, truck drivers on the interstates, factory workers, university administrators, and grant-making agencies that fund the research. The totality of these infrastructures and superstructures—humanity, in short—will never come to life on the screen.

When one decomposes culture into its constituent elements—identity, kinship, collective representations, appropriation of the body, reciprocity, creation of community, cognition, or any other cultural elements one wishes to add—there is no element on the list that cannot be reproduced with greater speed and accuracy *in silico*, via computer simulation. Cyborgs, synthetic holographic environments, virtual communities, and computer-generated art, music, and literature have joined artificial intelligence (AI) and arcade games as virtuoso displays of computational prowess. "Expert systems," for example, have been designed to solve tough, massively dimensional problems, whether weather forecasting, calculation of structural stresses, or the interpretation of patients' symptoms. Many of these in their own limited domains are faster, more accurate, and more satisfying than are human experts. Likewise, computer games reproduce the visible and aural images of battle, flight, or NFL football, and fine-tune a balance between challenge and mastery far more efficiently and safely than physical sports or actual combat.

They achieve this performance, of course, through reductionism, by stripping problems down to their core and ignoring peripheral issues that are not part of their programming. Thus, a medical diagnostics system might accurately conclude that a patient's obesity is related to hormonal imbalances and prescribe appropriate medication, without inquiring (unless programmed to do so) into the life circumstances that might have triggered the imbalances. A computer game can be set to offer just the right level of challenge to a player's dexterity or visual acuity, without the messiness of details such as recruiting other players, waiting for the beginning of spring, finding a playing field, or years of practice mastering a broad range of motor, conceptual, perceptual, and social skills. Such *decontextualizations,* as John Seely Brown and Paul Duguid note in *The Social Life of Information,* "represent forces that, unleashed by information technology, will break society down into its fundamental constituents, primarily individuals and information" (2002:22). Lost, of course, are the textures of connectedness that give information meaning and individuals a social life. When apparently aimless individuals are substituted for society, and apparently meaningless information for culture, we are clearly in a New World.

DESIGN AND DECOMPOSITION

Just as the character of human capabilities changes when augmented with technology, so too does the character of culture, that is, the role that culture plays in human adaptation. As classically conceived, culture was *the* source of human freedom, a mutually agreed-upon framework that freed us from the tyranny of biological passions and arbitrary coercion. In the absence of culture, humans would be slaves to their own desires for food, sex, and comfort, and their own fears of hunger, isolation, and intimidation. Our lives would be solitary, poor, nasty, brutish, and short. Culture as we know it has evolved over millennia of adaptation to natural and social environments. The latest stage in this evolution, I will suggest in this section, is the engineering of the habits of the heart to accommodate the imperatives of technology.

The beginnings of technology, whether ancient technology with the Sumerians and Egyptians, or modern technology with the Industrial Revolution, added a third dimension of adaptation, adaptation to a built environment. The built environment consisted of static structures (buildings, walls, roads), some of which survive even today, representing the intentions of monolithic powers such as states and empires. This allowed adaptation to be a gradual affair. The built environment represented the dead hand of the past, a historical continuity that gave ancient civilizations their timeless character. Although these civilizations did have their own dynamism, most typically in the form of wars and conquests, for them technology was a source of stability, not of disruptive change.

By contrast, the built environment of modern society is incessantly *changing!* In a technological society the normal learning rhythms of exploration and adjustment, common to all living systems, have become uncoupled from the bodily rhythms of birth, maturation, and death. They are likewise uncoupled from the agricultural rhythms of cultivating and harvesting, or the liturgical rhythms of celebration and mourning. Agricultural societies learn from one year to the next which crops and which methods of cultivation work best, and they make yearly adjustments based on the previous year's experience. Human families support and learn from previous generations' cycles of birth, growth, maturation, and senescence. In a technological society, by contrast, the challenges of learning are anchored only to technology. Learning how to program a VCR, for example, has no meaning more profound than mastering the gizmo itself.

A distinguishing characteristic of modern technology is its self-generating character, its compounding of existing capabilities with new devices, new methods, and new purposes. More precisely, industrial technology democratically empowers diverse groups—laboratories, universities, graduate students, militaries, start-up businesses, networks of friends—to create new, disruptive regimes of artifacts and imperative control to which wider circles must then adapt. Thomas Hughes' (1987) early industrial imagery of system building and forward salients, discussed in chapter 3, is now replaced by a postindustrial terrain of technological insurgency, where the smartest and agilest innovators use asymmetric warfare to topple industrial dinosaurs. An intuitive understanding of microcomputing for example, and not size or market dominance, enabled Microsoft to outmaneuver IBM and eventually dominate the personal computing industry, just as an agile Google is now outmaneuvering an aging Microsoft. When some graduate student achieves the right combination of social presence through ubiquitous computing, personal expression through styling, and service distribution through urban planning, the sun will begin setting on automotive transport as we now know it.

A substantial body of literature in management studies makes it clear that the traditional routines of administration once taught in business schools are no longer adequate for "thriving on chaos" (Peters 1987) or "riding the waves" of disruptive change. Classics such as Barnard's *The Functions of the Executive* (1946) or Drucker's *The Concept of the Corporation* (1946) from the industrial era offer very little guidance in an overpopulated world of global competitors, technological innovators, and restless customers.

In this turbulent environment, the most widely accepted strategy for adjusting humans to machines is to retrofit organizations to better conform to technological requirements. This is most evident in hazardous operational environments, where minor mistakes can quickly have major consequences. Several bodies of theory and practice have been created to better adapt organizations to hazardous technologies. For example, high reliability theory (HRT), developed by a group of sociologists, social psychologists, and engineers at the University of California, Berkeley, has identified a common set of characteristics in a variety of industries where hazardous operations and societal investments create a potential for catastrophic failure. From studies of three of these—naval aircraft carrier operations, a control room at a nuclear power plant, and air traffic control towers in the U.S.—the team of researchers identified several characteristics that contributed to the high reliability of these operations. Among these were:

- Shared understanding of mission, so that all personnel are focused on the same objective
- Redundancy of personnel and other resources, allowing for "multiple eyes" to watch for mishaps and multiple hands to intervene
- Constant training mode, in which all personnel are engaged in training either of a person below them, or in preparation for a more responsible position
- Accountability and responsibility pushed to the lowest levels of the organization, and high standards of accountability
- Open, nonhierarchical channels of communication, allowing problems to be communicated instantly without bureaucratic filtering (Roberts 1990; Roberts and Rousseau 1989; Weick and Roberts 1993)

These characteristics are supported by two other characteristics that make them difficult to generalize: First, redundant personnel and

The control room of the Davis-Besse Nuclear Power Station in Oak Harbor, Ohio. The power station was shut down for two years while the reactor head was replaced due to corrosion damage.

frequent training depend on budgets that cash-strapped governments and corporations may lack. City police departments, for example, have to choose between hiring additional police officers and training officers currently on the force, to improve public safety. Second, the sites that they examined—aircraft carriers, control towers, nuclear power plants—are all sites with strong institutional and sometimes physical boundaries. This enables managers to exert more consistent authority than is the case in places such as hospitals or construction sites where personnel freely wander in and out. The fact that settings with boundaries are either military facilities, government services, or regulated utilities make it difficult to generalize high reliability to private industries that resist government regulation and militaristic order.

High reliability theory emerged out of academic research. A practical discipline, Crew Resource Management (CRM), emerged out of the experience of aviation practitioners—flight crews—who discovered that single-pilot habits and traditions were inadequate for technologically advanced, multicrew airliners. CRM trains aircrew in cognitive and social skills, including goal setting and prioritization, task sharing, workload management, communication, situational awareness, decision making, and stress management. Effective CRM, for example, permitted Captain Al Haynes to work with two other pilots in landing a crippled DC-10 in Sioux City, Iowa, in 1989. All three pilots—captain, first officer, and a third pilot who happened to be on board—credited their ability to synchronize their efforts to their CRM training (Kern 2001).

CRM and HRT have in common a focus on the real-time logics of individuals at operational levels, whether pilots or deck hands. Efforts to generalize these methods to new contexts such as management decision making, where policy logics and corporate orders have greater force, have had less success. Social psychologist Karl Weick and coauthor Kathleen Sutcliffe, in *Managing the Unexpected* (2007), have extended the insights of HRT into the upper reaches of management, noting that shortfalls in open communication and situational awareness can do damage as much to corporations as to airliners. Weick and Sutcliffe give the example of how management crippled a merged Southern Pacific/Union Pacific Railway, overriding the local experience of operational personnel, bringing traffic to a halt, and costing the Texas economy approximately $4 billion. Organizational complexity extends far beyond the executive suite to include the dominant customers in an industry, labor unions that may have cooperative or hostile relationships with management, stock markets that whipsaw corporations, and government regulators that in various ideological regimes are tolerant or intolerant of industrial accidents, shareholder risk, or executive inconvenience. In HRT theory, high reliability starts at the top; in practice, however, when starting at the bottom, high reliability sometimes stalls one or two levels above the shop floor.

Both CRM and HRT are examples of "normative management" (Barley and Kunda 1988) in which the familiar methods of control— material incentives, direct orders, rewards and punishments—are supplemented by a manipulation of the moods and motivations of the employees. Gideon Kunda has described the steps taken by management at a New England technology company to create a corporate culture in which employees are motivated as much by their love of the company and its products as by the material incentives that it provides. In *Engineering Culture*, Kunda (1992) describes the ideology of management to encourage developers of technologically sophisticated devices toward open communication, creativity, and 80-hour workweeks.[3] As employees enact this ideology through staged rituals, stylized gestures, and personal displays, they face challenges, not unlike those in a MUD, of wondering just where is the authentic "self."

Among those who operate sophisticated technologies there are similar forms of normative management, often captured in the slogan "safety culture." Safety culture consists of such unexceptionable traits as senior management commitment to safety, shared care and concern for hazards and solicitude over their impacts on people, realistic and flexible norms and rules about hazards, and continual reflection on practice through monitoring, analysis, and feedback systems (Pidgeon 1997:7). This "culture" is clearly prescriptive, rather than emergent, and has the potential to reduce "culture" to one more item on a safety checklist. Like "engineering culture," it is a retrofit of the habits of the heart to serve managerial imperatives.

Perhaps the best known effort to retrofit culture to the demands of industrial technology is that of the Dutch social psychologist Geert Hofstede. In the 1970s Hofstede conducted a study of attitudinal differences among IBM employees in 40 countries around the world; this study yielded a dimensional decomposition of culture that has been widely cited in management literature ever since. Hofstede's work is an effort to understand "irrational" motivations within regimes of instrumental rationality. These regimes of instrumental rationality—modern organizations—are the context for modern technology, either in its production, its deployment, or its regulation (Hofstede 1980).

To understand the motivations that different nationalities bring to modern organizations, Hofstede administered a questionnaire consisting of 158 items to middle-management employees of IBM. Questions on the survey included such matters as, "One can be a good manager without having precise answers to most of the questions that subordinates may raise about their work (Strongly agree, agree, neutral, disagree, strongly disagree)," or "In order to have efficient work relationships, it is often necessary to bypass the hierarchical lines." Hofstede then analyzed the results and discovered four "factors" that he posits are the fundamental dimensions of culture: (1) individualism vs. group orientation, (2) masculinity vs. femininity, (3) power distance, and (4) uncertainty avoidance.

Although Hofstede's method is, from an anthropological point of view, deeply flawed (suffering from, among other matters, ahistoricity, biased sampling, ethnocentrism, lack of nuance, reification, and spurious precision), it has achieved wide credibility within managerial studies of the influence of culture on technology-intensive industries. This credibility comes from the fact that, for Hofstede (and those who cite him), culture is no longer an integrated totality, but rather a storehouse filled with collective attitudes cut from similar parts templates. Thus, in terms of approaches to education, Confucian respect for learning and American pragmatism can be mapped onto the same scale of power distance (68 and 40, respectively), suggesting that the dials and knobs of culture can then be adjusted to meet the demands of modern industry.[4]

In sum, faced with demanding technologies, whether nuclear power plants, automated flight decks, or air traffic control centers, human irrationality is brought under control by disassembling culture into attitudes, behavior patterns, motivations, stylistic gestures, and values, and then re-assembling it into totalities such as safety culture, Crew Resource Management, or high reliability. In the global spread of technology, these techniques have become a critical management discipline: a switchology of culture.

THE SECRET LIFE OF THE HUMAN FACTOR

One characteristic of technology is its restlessness, its continual seeking of new configurations, new problems, new contexts, and new uses. This "technological exuberance," as I characterized it in chapter 1, means that technology never sits still, never assumes a stable configuration. This creates a dilemma. From an engineering perspective, "the human factor" has too many degrees of freedom; it must be brought under control through *technê* such as cultural surveys, Crew Resource Management, safety culture, or high reliability. Yet, none of these schemes, perfect as they are, can anticipate every imperfection in the world, whether the external world of operational environments or the internal worlds of the human heart. The only human factor that can safely sustain complex, hazardous technologies is the trained, insightful, experienced, and creative operator whose requisite unruliness matches the unruliness of the technology.

A path-breaking study of these dilemmas of "cultures of control" is provided by a study of operations and management in the nuclear power industry by Constance Perin (2005), an anthropologist at the Massachusetts Institute of Technology. Perin compares the differential orders of engineering, corporate policy, and regulation as creating an array of "trade-off quandaries," as operators and managers struggle daily to produce electricity, maintain safety, and earn a reasonable profit. She then contrasts these regimes of "command and control," a standard engineered order, with the operators' experienced-based strategies of tacit knowledge, informal rules, real-time logics, and workarounds as they maintain operation and safety in a highly complex and highly dangerous operation.

Nuclear power is a technology that is at once far more hazardous and far more complex than its precursor in coal-fired power plants. The objectives of confining a nuclear reaction also dictate that the system will be tightly coupled—so much so that a nuclear plant, Three Mile Island, furnished the type case for Perrow's (1984) analysis of "normal accidents," in which complexity and tight coupling conspire to make mishaps "normal." The Armageddon of a potential nuclear catastrophe places this technology at the limit of human capabilities. Nuclear power's hazards and complexity make nuclear safety an ideal site for examining the potentials and limits of human ingenuity.

Based on years of ethnographic observation and study within the industry, Perin (2005) showcases events and incidents where the front-line operators must improvise their own real-time logics, at times covertly and at times as an alternative to the standard procedures. The standard procedures did not reveal a stem-disc separation in a valve in the containment structure, confusing labels on circuit breakers that led to a reactor trip, or a potentially life-endangering voltage leak in a step-up transformer. Operators successfully coped with these incidents by applying their own context-based experience. In both the public understanding of technology and the hierarchy of knowledge within technological communities, experience, insight, real-time logic, and intuition are trumped by procedures, numerical indicators, policy logics, and managerial authority. The only way that the operators are able to "shoulder the risks" of safe operation are to supplement or at times work around the formal procedures with their own informal, unauthorized improvisations.

Perin adds a fine-grained, anthropological "I was there" authority to generalizations that others have made, namely that the realities of technology are far more contingent and improvised than the public image of highly engineered, highly predictable devices. Plans are created not so much to guide practice as to make sense out of improvisation (Suchman 1987). Deviation from standard operating procedures (SOPs) is a frequent practice (Vaughan 1996) in that informal rules provide more immediate guidance than manuals or procedures. *Yet, the public acceptance of technology, and the autonomy granted to technological activities, is based on a contrary assumption that the experts know what they are doing in following procedures rather than making them up as they go along* (Wynne 1988).

Nuclear power may prove the limits of technology in a democratic society. An industry whose power is measured less in the billions of watts it produces than in the billions of lives it potentially touches is able to produce electricity, maintain safety, and earn a profit only with a culture of compromise whose burdens fall primarily on those on the lowest rung: the operators. This culture of compromise flies in the face of the other orders of the industry: the "parts template" of engineered order, the policy order of the investors, and the regulatory order of the Nuclear Regulatory Commission. No one of these orders by itself can design, operate, and maintain a nuclear power plant; yet, the culture of compromise, lacking the legitimacy of these orders, can only do its work backstage, *sub rosa*, through workarounds, informal arrangements, or a *bricolage* of informal adaptations.

The public acceptance of technology is the ultimate human factors challenge. In the earliest days of industrial technology, improved productivity led to public acceptance, and externalities such as pollution of waterways, disruption of the idylls of rural life, and the immiseration of labor could be considered unfortunate, out-of-sight by-products. Technologies were relatively simple and did not require massive institutional superstructures. As technologies became more powerful, both energetically and aesthetically, public support was sustained by the promise of democratic prosperity and the association of technology with

sublime and awesome magical powers. That these powers could sometimes cause local damage, whether in an airplane crash, a chemical
plant explosion, or bacterial contamination in a food-processing plant,
rarely challenged the faith that the public placed in technology.

Nuclear power, however, is a technological singularity both in its
unequalled destructive potential and in its contrast between its image
and its operational realities. There is no technology of greater destructive
potential that could possibly serve any social purpose.[5] As the all-too-
human operators, managers, and maintainers shoulder its ongoing risks
of nuclear and electrical safety, they also hazard the risks of massive public rejection of their technology, their industry, and their achievements.

Public acceptance of an imperative order, whether of a government, a corporation, or a technology, is a problem that Max Weber identified as a problem of legitimation. Legitimation most broadly rests on
the cultural values of the society in which the order exists—not ideolog-
ical or degenerate forms of culture, such as corporate culture, but a culture that is most broadly shared and accepted within the society. Yet,
today, we find that the producers and custodians of cultural values of
our industrial society are divorced, alienated, or aloof from those who
create and operate their perfect infrastructure. When the producers of
culture are separated, like Athenian aristocrats, from the producers of
material goods, then the society is uneasily poised on a precipice not of
collapse but of substantial cultural or technological regression, or more
likely both.

Notes

[1] Adequate evidence for this comes from the fact that the one coherent institution for providing full-spectrum medical services, the Veterans Administration, is far ahead of the
fragmented and chaotic private sector in adopting systems of electronic medical records.

[2] Readers whose only experience with computers is with boxes sitting on top of desks, or
notebooks tucked inside briefcases, may not recall that as recently as the 1980s a "real"
computer was a mainframe: A large installation, sitting inside a climate-controlled
room, with rows of tape storage devices and card-reader inputs, and tended by white-
coated technicians. The advent of the personal computer democratized the information
processing that was formerly accessible only to an elite.

[3] Within the software industry, there is a wry comment, "You have a lot of freedom: you
can work any 80 hours in the week that you wish." This comment captures the ambigu-
ities and ironies of normative management.

[4] To be fair, in a subsequent book, *Cultures and Organizations,* Hofstede does supply historical and philosophical contexts for some of his earlier observations. The core of his
work, though, is still the 158-item survey.

[5] Knowledgeable readers might suggest that recent experiments with plasma physics,
which promise unlimited cheap energy, might disprove this *nec plus ultra* (nothing further beyond) of nuclear power. A passing acquaintance with climatology or with the second law of thermodynamics should cause one to doubt the benefits of unlimited energy.

Chapter Six

Technoscience versus Culture

Who owns culture? Within any community there are gradations of authority, not only over the members of the community but also over their hearts and minds. Although the incumbency and legitimacy of authority might be questioned, and dissident groups may rebel, the existence of authority, both sacred and secular, is not. In pre-industrial state societies such as Rome and Greece and Medieval Europe, academic and religious gatherings and institutions were the custodians of Truth. Kings and emperors might advance questionable claims to sovereignty, but they sought or hired priests or scholars, the learned authorities on all matters, to legitimate their claims. Peasants might be terrified by earthquakes or thunderstorms, but the village priest could explain that it was all part of God's design.

With the Scientific and Industrial Revolutions the authority of priests and scholars began to decline. Priests who understood heaven had little comprehension of the heavenly facts that Galileo's telescope revealed. Scholars knowledgeable in the law and ancient texts had little to say about the new laws that experimental scientists were discovering. By the middle of the 20th century, this erosion of traditional intellectual authority was complete, although most intellectuals would require another generation to catch on to the fact.

An early notice of the breakdown of intellectual authority was a lecture, "The Two Cultures," given at Cambridge in 1959, by C. P. Snow, a British novelist and scientist. In his Rede Lecture, Snow described the lack of communication between the sciences and the humanities: Many scientists, Snow observed, have not read Dickens, but neither could many humanists describe the Second Law of Thermodynamics, or even supply a workable definition of "acceleration." The lecture provoked a broad reaction, ranging from common applause to donnish hysteria, confirming that Snow had tapped into some deep-seated anxieties among the literate public.

In this chapter, I present the "Two Cultures" controversy as a microcosm of issues of cultural and technological authority within postindustrial civilization. The question of who should set standards, broadly speaking, for a society, which had seemed settled for Matthew Arnold in the 19th century, was completely upended by the middle of the 20th. After reviewing the issues that Snow raised and the reaction to them, I examine how "culture" and "science" have become contested terrain. "Science" and "technology" are equated within popular and even academic discourse. This equivocation obscures the fact that technology's truth claims are based primarily on imperative order ("standards") rather than efficient instrumentality, which is always context-dependent, that is, contingent. "Science and technology," or "technoscience" in academic discourse, has thus become a master trope, a key symbol for authority in postindustrial society. Science and technology authoritatively speak to a broad range of issues in health (medical technology), economic well-being (improved industrial technology), law enforcement (surveillance technology), and, of course, national defense.

STRANGERS AND OTHERS

Closed intellectual circles at times generate closed controversies whose spillover into public discourse takes all by surprise. On such occasions, a confusion reigns over the legitimacy of arguments and the legitimacy of those who are offering them. Outsiders' viewpoints are dismissed *ad hominem*, and social exclusion obscures weak conceptual foundations.

From a perspective of 50 years, the controversy generated by C. P. Snow's Rede Lecture, "The Two Cultures," is as much about identity and authority as it is about knowledge or inquiry. The author of "The Two Cultures" was the son of a clerk in a shoe factory, a tradesman who was also a distinguished organist. Snow himself, a physicist, novelist, and public official, observed that the communities of natural scientists and those of "literary intellectuals" co-existed in mutual incomprehension, with neither understanding the basic tenets or purposes of the other. Each was a stranger to the other. The literary intellectuals were men (rarely women) of social distinction in the tradition of Matthew Arnold, fellows at Cambridge and Oxford, occupying the commanding heights of intellectual life in Britain. The scientists, in the tradition of Cavendish and Rutherford, were experimentalists down in the laboratories discovering the Laws of Nature that propelled Britain's Indus-

The Clock Tower, Trinity College, Cambridge University.

trial Revolution. This fissure in cultural authority—"anarchy" would have been Matthew Arnold's term for it—seemed to Snow a harbinger of a more general decline.

Snow used the term "culture" in two senses. A traditional sense of culture "intellectual development, development of the mind," echoes Matthew Arnold's formulation in *Culture and Anarchy*:

> a pursuit of our total perfection by means of getting to know, on all the matters which most concern us, the best which has been thought and said in the world, and, through this knowledge, turning a stream of fresh and free thought upon our stock notions and habits. (Arnold 1994 [1869]:5)

This traditional sense of culture as cultivation that liberates the mind from narrow prejudices carries with it obvious marks of class privilege. A second sense, which Snow embraced, was the "technical" sense: "[Culture] is used by anthropologists to denote a group of persons living in the same environment, linked by common habits, common assumptions, a common way of life" (Snow 1998[1959]:64). Although today's anthropologists might be more careful about distinguishing between groups on the one hand and their common habits and assumptions on the other, this understanding of culture as a way of life lent force to Snow's argument.

Two learned communities were notably on the fringe of Snow's bifurcation. Snow took an interest in social historians and related disciplines (sociology, demography, political science, economics, psychology), and suggested that they might constitute a "third culture" (1998[1959]: 70), although he never developed the point, perhaps because he lacked the up close familiarity with them that he had with both novelists and natural scientists. On the other hand, technologists and engineers apparently did not constitute or have a culture. They were not admitted to the High Tables. "Pure scientists have by and large been dim-witted about engineers and applied scientists" (32) as well as "devastatingly ignorant of productive industry" (31). Although applied scientists might share in the dignity of Science, engineers, in a common class prejudice of the time, were little more than tradesmen. (Davenport 1970:11–16). Four years later, in "Two Cultures, a Second Look" Snow softened the "clear line between science and technology (which is tending to become a pejorative word)" (67); nevertheless, the view of "technology" as applied science mucking around in industry, and the class prejudices this reinforced, remained.

Snow (1998[1959])identified not only "common attitudes, common standards and patterns of behaviour, common approaches and assumptions" (9) for the scientists but also a common moral purpose notably lacking for the literary intellectuals: saving the world's poor from starvation. "Industrialisation is the only hope of the poor," an enterprise in which scientists (pure and applied) were at the vanguard. Snow had nothing but contempt, which would soon be returned in kind, for academics whose comfortable, well-endowed chairs enabled them to be disdainful of industry:

> It is all very well for us, sitting pretty, to think that material standards of living don't matter that much. It is all very well for one, as a personal choice, to reject industrialisation—do a modern Walden, if you like, and if you go without much food, see most of your children die in infancy, despise the comforts of literacy, accept twenty years off your own life, then I respect you for the strength of your aesthetic revulsion. But I don't respect you in the slightest if, even passively, you try to impose the same choice on others who are not free to choose. (25–26)

An intense and at times vicious reaction (Leavis 1962; recounted in Cornelius and St. Vincent 1964; Davenport 1970) that reverberated even 36 years later (Brockman 1995) suggested that Snow had articulated a widely held anxiety over the state of public discourse. Most notoriously the critic F. R. Leavis characterized Snow's lecture as having "an utter lack of intellectual distinction and an embarrassing vulgarity of style," and Snow's place in the 20th century is "characterized not by insight and spiritual energy, but by blindness, unconsciousness, and automatism. He doesn't know what he means, and doesn't know he doesn't know." (Leavis 1962:10).

A more even-tempered assessment of this controversy, from the perspective of the other side of the Atlantic, was supplied by the critic Lionel Trilling (1962), who placed "culture" in the context of other modes of collective activity: "The concept of culture affords those who use it a sense of the liberation of their thought, for they deal less with abstractions and mere objects, more with the momentous actualities of human feelings as these shape the conditions of the human community, as they make and as they indicate the quality of man's existence." Yet, Trilling wonders whether culture, "in an age dominated by advertising," is debasing identity to a matter of style in the cut of one's trousers (1962:474–475).

Leavis' outburst suggests that Snow had articulated a thought that the critics found embarrassing but impossible to refute: that the scientists were actually an ascendant intellectual class. The scientists were this despite being an Other, aliens to the literary intellectuals who found the scientists characterized by incomprehensibility, strangeness, and questionable manners. A proud community long regarded as the custodians of the best which had been thought and said, was, in Snow's reckoning, toppling into the dustbin of intellectual history.

UNDERSTANDING CULTURE

Discussions of "two cultures" or a "third culture" or "technology and culture" (Brockman 1995; Davenport 1970) betray almost no awareness that they follow a hundred years of empirical investigation into the cultures of the world, seemingly content with the idea that within the libraries of Europe and North America one can learn all there is to know about culture. Culture, in this view, following Matthew Arnold, is something that separates the best from the rest, and to speak the cultures of Pygmies of the rain forest, or Native Americans from the Great Plains, or Eskimo from above the Arctic Circle, really is nonsensical.

Whether the cultures of the world merit empirical attention is an issue that is as old as empires. Among the Greeks, Thucydides considered the customs of barbarians unworthy of notice, although Herodotus took a keener interest. As a profession, anthropology was founded on an affirmative answer to this question, and has dispatched legions of Herodotuses to every corner of the world to investigate other peoples' ways of life—from the dietary habits of !Kung to the sleeping arrangements of the Siriono or to the sophisticated toolkits of the Inuit. From such investigations, we extend our understanding of the range of human possibility and better comprehend why "irrational" phenomena such as religious fanaticism are vexing problems, even in our enlightened age. These small-scale phenomena of native cultures or developing nations might seem to the members of developed nations to be the marginalia of human existence. Subtle cultural gestures, however, create the strength and resilience of most cultures, and are ignored at one's peril. For anthropology, empathetic understanding of other cultures links a comprehension of human variety with an understanding of a common humanity. In an era of globalization there is perhaps no more urgent enterprise.

Cultures supply frameworks for meaning, for making sense out of the world, and for understanding what to do about it. These frameworks become so commonsensical that people reflect on them no more than fish reflect on the water in which they swim. The *fact* that in the United States, individuals are expected to take responsibility for their actions, whereas in other lands responsibility is assigned to family, to the gods, or to the state, is a cultural understanding. Likewise, the *fact* that in Germany businesspeople expect punctuality but in other nations exact punctuality is considered mildly offensive is not just a refinement of etiquette; it is a statement about who *we* are. Beliefs and values such as these are encapsulated in symbolic expressions whose logical coherence forms a cultural *system*. Cultural systems have a logical structure, such as that which David Schneider (1968), in an American version of French structuralism, describes for American kinship (the distinction between relatives in nature [that is, blood relatives] and relatives in law [by marriage or adoption]). Cultural systems, in the structuralists' view, have a logical (or analogical) order to them, in that they are organized around symbolic distinctions. How this cultural system is structured, and how it relates to historical events, is the subject of extensive analysis; the *fact* of its existence, however, cannot be questioned, even if, as Appadurai notes (1996:12), it is more comfortable to talk about "cultural" aspects of phenomena than about a reified "culture."

An important part of a cultural system is the hierarchy of knowledge that it establishes. In different societies, at different times, mysticism, empiricism, rationalism, or dogmatism have had the strongest claims to Truth. In the Middle Ages, the Bishop of Rome was a final authority, even on celestial mechanics. In contemporary Western societies, science is the preeminent way of knowing: Even those who deny science's conclusions, questioning Darwinian evolution, feel compelled to dress up their arguments as "creation science" or "intelligent design." In so doing they acknowledge the primacy of science and intellect, rather than resorting to an earlier dogmatism of "the Bible says so." The characterization of any conclusion as "scientific" automatically elevates its credibility.

In *The Coming of Post-Industrial Society,* the sociologist Daniel Bell (1999[1973]) identified the "knowledge class" as the leading class in contemporary society, having displaced the industrialists of the modern economy, who in turn displaced the landed aristocracy of pre-industrial states. Within this knowledge class Bell included scientists (natural and social), engineers, administrators (sometimes styled as "technocrats," inasmuch as their authority derived from technical expertise), and creative workers. Bell identified this class as central to the continued prosperity of a postindustrial economy, although he noted that in the United States, most research and development expenditures (the lifeblood of the knowledge class) went toward defense and space research. Bell followed, by nearly 20 years, Jacques Ellul's more pessimistic appreciation of the same facts, in which Ellul observed that "technique" (I would say managerial rationality) was becoming all-pervasive in the modern world (Ellul 1964, from *La Technique* [1954]).

From a cultural perspective, David Hess in *Science and Technology in a Multicultural World* interrogates the foundations of Western "technoscience," demonstrating that many of its core assumptions are rooted in the social distinctions and cultural assumptions of the societies that produced it (Hess 1995). Although the natural sciences are often thought of being above or outside cultural influences (the gravitational constant, after all, is the same on both sides of the Pyrenees), Hess provides numerous historical examples of how national cultures, religious motivations, and racial prejudice shaped the trajectory of scientific discoveries. For example, a controversy in the early 20th century

IBM's Thomas J. Watson Research Center (designed by the Finnish architect Eero Saarinen) exemplifies the corporate laboratories that were at the center of Daniel Bell's knowledge class in the years after World War II.

over statistical tests (Pearson's r vs. the Q test) had a close relationship to the class agendas (eugenics vs. the maintenance of privilege) of their respective proponents. Those advocating for a statistical test appropriate to normal, continuous distributions (where differences are a matter of shading, or degree), wanted to use its results to further "a program that supported selective human breeding in order to improve the population" (Hess 1995:24f.), whereas those who wished to maintain class privilege wanted more clear-cut (i.e., nominal or categorical) tests.

A standard narrative of scientific progress, in our enlightened age, dismisses such cultural contaminations of science as self-correcting. Science advances through the refutation of such errors. Yet, even today, it is easy to spot uncorrected "scientific" conclusions that rest on dubious social assumptions. Paul Rabinow, for example, in his chapter, "Galton's Regret: Of Types and Individuals" (1996), demonstrates that DNA identification, a forensic technique with impeccable scientific authority, assumes the boundedness and precision of populations such as "Italian Americans," "Asians," "African Americans," and "Hispanics" at the very time that these categories are collapsing into chaos. Statements of the sort that certain alleles, which are the basis of DNA identification, are 50 times more typical of population X (Asian Americans, for example) than for population Y ("Hispanics"), have an empirical imprecision that is masked by arithmetical exactness and "scientific" authority.

An important conclusion that can be drawn from Rabinow's example is the widespread confusion of science, technology, and technique. DNA identification is a *technique*—an orchestration of multiple faculties to solve a specific problem. Its standardization (Kirby 1990; Office of Technology Assessment [OTA] 1990) justifies calling it a *technology*. Yet, aside from its methodical use of knowledge derived from biologists and chemists, it has nothing to do with science. One might as well call driving an automobile "scientific," inasmuch as it employs Newton's laws of motion. "Scientific," in this everyday parlance, simply means "methodical," which could be (and has been) applied to almost any activity, carefully and insightfully pursued, perhaps with some quantification thrown in: scientific cooking or scientific housecleaning, for example. *Science* is the rigorous and systematic discovery of underlying regularities in nature and humanity, such as the molecular structure of deoxyribonucleic acid (James D. Watson and Francis Crick suggested the first model of DNA structure) or the connection between segmentary identities and predatory expansion (Sahlins 1961). *Applied science* uses such results to solve practical problems, such as the design of electromechanical systems, the identification of corpses or culprits, or the formation of political alliances. The actual *use* of such methods involves considerable technique, including the collection and storage of samples, the calibration of instruments, and the interpretation of results. However, to enlist these techniques in legal proceedings, with the authority of the government behind it, requires that they be standardized: turned into technology. The misuse of a folk sociology (the "racial" categorizations of the United States, *circa* the 1990s) makes it clear that this technology rests on a foundation of preexisting cultural assumptions. This confusion of science and technology, reflected in the academic term "technoscience" and embodied in numerous public programs supporting "science and technology," makes it clear that even "science" has become totemic: the ultimate technototem.

This totemic status of science should not be surprising, given the prestige of science in contemporary Western culture. For approximately 300 years, the West has distinguished itself from the rest by constructing an original relationship with the universe, an original relationship based on the findings of experimental science. This relationship with Nature is fundamental to Western identity. Culture, likewise, is closely related to identity. Cultural identities are constructed in relationship with objects whose identity is nonnegotiable: fish and fowl in some tribal societies, the Laws of Nature in ours. Different groups around the world to different degrees conceive of themselves as having a distinct identity—a set of symbols, affiliations, separations, and norms that define who they are as a distinctive people. Totemic identities, being nonnegotiable, carry an implication of exclusivity. Although, as a matter of personal identity, one can be a mother, a business executive, and a community leader, as a matter of totemic identity, one cannot be a police officer *and* a firefighter, any more than an Ojibway can be a member of both the Eagle Clan and the Wolf Clan.

The degree of openness or closedness of a system of meaning and identity raises an important point for empirical study: how readily can an outsider understand another culture? Anthropology has long since outgrown its imperial past, in which a Margaret Mead could go to the South Seas for two years and return as an authority on All Things Samoan. Anthropologists now approach their subjects, whether Pacific islanders or natives closer to home, with a great deal more caution, modesty, and reflexivity than was the case before nonanthropologists pointed out some of the arrogance and factual inaccuracies in the traditional approach. Yet despite this, anthropology is still respected for the up-close engagement it brings to small-scale subjects: for its unique ability to make the foreign familiar, and to see familiar objects from a fresh perspective.

Anthropology's focus on small-scale societies poses a problem for the anthropological understanding of technology. Technology, as I have defined the term (instrumentalities involving authoritative standards encoded within independent representations, creating a zone of autonomy) is associated with a state-level of social order and comes to its fullest flower within large-scale civilizations. Unless one equates "technology" with "tools" (a misapprehension discussed and laid to rest, I hope, in chapter 2), one needs additional research apparatus beyond ethnography, as indispensable as it is, to understand the cultural contexts of technology. The first of these contexts, of course, is the source of authority invested in technology.

AUTHORITY

In an earlier chapter, I observed that technologies are projects of civilizations. For different eras, the archetypal technologies took their character from the sort of civilization that produced them: even as late as 1823, "technology" still referred primarily to the study of static structures (Bigelow 1829). Only further into the Industrial Revolution did machinery come to be seen as the highest form of technology, even though machines had been around for millennia. By 1954 Jacques Ellul could note the widespread confusion of machinery and *la technique*. In a postindustrial age, computation and communication devices are considered in common parlance the truest expression of technology, whereas the "technological" character of "low-tech," clunky devices, such as locomotives, is suspect.

From the perspective of archaeology, a civilization is characterized by a state form of government, writing, agriculture, cultural complexity including a class system and functional specialization, and monumental architecture (Steward 1955). With this understanding of civilization, we can see that technology in some form is a core element of civilization; just as the monumental architecture through which state authority projects itself requires technological sophistication, so, too, does agriculture (and later industry) have a technological base. Writing is essential to the standards that distinguish technologies from local tools, just as cultural complexity necessitates that technology achieve a degree of autonomy from the social order through standards. Even if every farmer uses a different form of plow, some standardization of weights and measures is required by the state's tax collectors. These matters were discussed in chapters 1 and 2. Technology is an embodiment of state authority, either directly through military technology, infrastructure projects, or monuments, or indirectly through the enforcement of standards. The enhancements in productivity achieved by common standards is consistent with those achieved by knitting a nation together with a network of roads.

The manner in which states embrace (contemporary United States), accept (early modern France), demean (Athens), ignore (pre-unification Germany), or constrain (Ming dynasty China) technological development has a strong determination on the place of technology within the society. States include not only formal public entities but also those that are chartered as extensions of state power, joint-stock corporations, for example. Without the visible hand of the state, technology does not exist. Although a lone inventor such as an Edison or a Siemens might invent a clever new device, the development of the device into a true technology (implying standards, social integration, and accepted purpose) depends on broad-scale authoritative institutions.

When we examine technology, therefore, one question we are asking is, what state purposes does the technology promote? Within all civilizations, technology has been concerned with production and distribution, the projection of state power (both internally and externally), and the maintenance of order. Thus, the roads of China in the Ming dynasty were not built to promote tourism among the Chinese people, but instead they were constructed for the emperor's bannermen to relay messages to the far provinces. Likewise, the Roman roads of Gaul were primarily military roads.

The technological monuments of the 1930s, 1940s, and 1950s in the U.S. were nation-building projects: the Tennessee Valley Authority, the Rural Electrification Administration, the building of a network of air transport, or the Interstate Highway System were all projects that stitched together the American nation into a more perfect union. The growth of broadcast media, and the rise of mass consumption, could also be added to this list of nation-building technological achievements, similar to the construction of "imagined communities" in an earlier era of print capitalism (Anderson 1983).

After World War II, the growth of technology in the United States replaced nation-building with empire-building. The aerospace industry was the first harbinger of this, but the space program, launched in earnest in the late 1950s, was pure competition between an expanding Soviet empire and a reluctant American empire. Technology of imperial reach in aerospace, and of corporate control in IT (Beniger 1986), replaced technology of social benefit as objects of dominant interest. During the New Deal, the federal government made substantial investments in technologies for improving the lives of farmers, improving schools, and improving transportation. This was done not out of charitable impulses but in order to build the America nation. By the 1970s, the federal government's interest turned away from nation-building toward empire-building, and technological emphasis shifted accordingly.

By the 1950s, the alliance between the state and science (and technology) was firmly in place, although what sort of New Man (the scientist) or Old Boy (the literary intellectual) should occupy its positions of authority was still a raw issue. By the time that Daniel Bell wrote in the early 1970s the question was settled: scientific and engineering authority were looked to by all for solutions to all matters great and small, including industrial productivity, national defense, health, poverty, crime, the environment, as well as earthquakes, floods, and thunderstorms. Until the reactor accidents at Three Mile Island (1979) and Chernobyl (1986), technology's authority was questioned only by nervous intellectuals such as Rachel Carson or Paul Ehrlich. If Snow were merely discussing epistemology or the sociology of academic life or the reading habits of physicists, he would hardly have triggered such an intense reaction. The panic of the literary intellectuals such as Leavis was a panic over their looming irrelevance. The disciplines (literature, ancient languages, fine arts, natural and moral philosophy) once considered indispensable for equipping gentlemen for public service were being elbowed aside by those of scientists, engineers, technologists, and vulgar tradesmen.

BOUNDARY EXPLORATIONS

The conceptions of culture that we have been examining thus far have assumed an idealist understanding of culture, very much in both traditions of "a learned system of shared understandings" and "the best which has been thought and said." Material objects, including much of what most people think of as technology, at best fit awkwardly within this idealist conception of culture. This question of culture, technology, and materiality is one of the guiding questions I raised in chapter 1.

An alternative perspective sees both technology *and* culture, contrary to Matthew Arnold, C. P. Snow, Lionel Trilling, and even my teacher on these matters, David Schneider, not as mental abstractions but as sensuous constructions of physical objects and experiences (Pfaffenberger 1988b, for example). This perspective, developed in studies of material culture yet extensible to the totality of culture, sees the investment of meanings, such as those attached to gender and life history, in material objects not as an abstract intellectual exercise but as an active engagement between people and the material things with which they build their lives. To anyone unfamiliar with the 1950s, with the Beach Boys, or with automotive history, the "real fine 409" is utterly incomprehensible: a curiosity of a forgotten era. To those who were there, "She's real fine, my 409" (a 1959 Chevrolet Impala coupe) is a song with overtones of gender identity, teenage longing, and sensual appreciation, that, like Grant McCracken's essay, "When Cars Could Fly" (2005), provides a tangible image of how an object can be freighted with deep cultural meaning. Other forms of meaning invested in objects include fetishism, commodity, and companionship. Likewise tangible and sensory engagement with the worlds of people and the world of Nature (itself a cultural construction) locates culture within the practical activity of men and women.

Understanding the meaning of objects requires a close-in familiarity with both the object and those people who are invested in it. This familiarity is supplied by anthropology's research technique of ethnography. Ethnography is based on the assumption that one cannot understand a way of life without getting up close and personal—talking to people, living among them, sharing their experiences. In exactly the same spirit, I would suggest, one cannot understand a technology until one gets intimately involved with it—*not* as a user, but as a designer, a manufacturer, or a maintainer.

A notable characteristic of modern technology in its mature stages is that it seeks to hide as much of its functionality as possible from the user, exactly as some societies hide their most intimate customs from outsiders. Thus, automobiles today no longer have chokes, engine instruments, primers, or manual transmissions (mostly), and today's personal computers no longer require the user-supplied configuration of an earlier generation. The spirit of modern technology is to create a sense of miraculousness by hiding functionality behind smooth, shiny surfaces.

Only by becoming closely involved with the device on multiple levels and trying multiple options to design, make, or maintain it, can one discover an object's social boundaries. When one is up close and personal to technological objects, a new dimension of technology, not found in simpler tools and rarely mentioned in the social literature of technology, comes into focus. This is the technology developmental cycle: as a device, whether a software program or a domestic appliance, emerges into public view, it does through a set of definable stages—analysis, design, engineering development, production, commercialization, sustainment, and finally retirement. Analysis is the identification and definition of the problem, design is the creation of a solution, engineering development is the translation of the solution into something buildable, whereas sustainment is the continued support of the object and retirement is its eventual disposal. Not only do individual technological devices go through this cycle, but entire classes of devices also repeat it. This is a fact well recognized in the engineering literature (Boehm 1981) and the literature of environmental design (Hundal 2002), yet mostly overlooked in the science and technology studies (STS) literature (Perin 2005 is an exception).

Two important social consequences follow from this life cycle. First, the object is socially redefined, and redefined again, as it goes through the cycle: the prototype that looks like a shiny new toy to the engineers developing it grows into a status symbol for the early adopters, matures into an appliance for the consumers using it, and figures as a business calculation for the executives selling it. As the device ages, it becomes a burden for operators whose use of it is fully rationalized with production quotas and skimped support. It also becomes a headache for the wrench-turners who must maintain it with out-of-date manuals and difficult-to-find replacement parts. Finally, it becomes a curio for collectors or a planning nightmare for the municipal authorities who must dispose of it.

Although the above description is drastically oversimplified, it demonstrates the life cycle that every technology goes through: flintlock muskets, that equipped Lee's army at Gettysburg, have gone completely out of style and, if preserved at all, are found only above fireplaces or in the prop-shops of civil war reenactors. Personal trucks, similarly, a status symbol in the 1990s, with rising fuel prices are now on the downslope of acceptance and will soon be cluttering junkyards on the fringes of most cities. Technologies go through this cycle at different rates. Some shoot through the life cycle at Internet speed, with the disposal of obsolescent five-year-old computers a problem for many municipalities that no one anticipated when the personal computer was a shiny new toy. Others traverse it more slowly: space capsules and jet packs are still in the early adopter stage, as they have been for more than 40 years.

A second social consequence of the technological life cycle is that it affects the cost-benefit calculus that usually is used to justify a technology. If one focuses only on a technology in its operational phase, one overlooks the subsidies that got it there, the infrastructure (including user skills) that keeps it there, and the externalities of its operation that eventually will have to be paid by someone. Externalities, by definition, are not accounted for, but they are nevertheless real costs. In the United States, with seemingly endless empty acres, old automobiles are dumped into the environment: the costs of disposal are externalized, just as the costs of infrastructure are subsidized. In many European countries, by contrast, purchases of new automobiles are taxed to include a fee to cover disposal costs, gasoline is taxed more heavily, and the retail price of fuel more accurately reflects infrastructure costs. Not surprisingly, in Europe, cars are smaller and more expensive, neighborhoods are more compact, bridges are maintained, and driving is not the default option for most errands.

Subsidies such as these are a demonstration of the importance of state support for technology. These subsidies come in the form of DARPA (Defense Advanced Research Projects Agency) developing the Internet, gasoline taxes funding highway construction, the Air Force funding much of the development of flight technology, or the FCC providing a stable environment for broadcast communication. Likewise, the acceptance of externalities such as degraded working conditions in the early years of industrialization, or pollution and climatic instability in the later years, are technological costs borne by the public, or at least by its less fortunate segments.

When a scholar has grease under his fingernails or acid burns on her lab coat or a smudged field notebook tucked under his arm, then he or she begins to see these subsidies and externalities, and the distance between the two cultures begins to decline. Each of these—the grease, the acid burn, or the dirty notebook—is a token of bodily engagement with the subject. Each of these indicates that the scholar has confronted his or her world of facts, whether mechanical devices, chemical reactions, or village life—with all five senses and a personal vulnerability that intensifies the experience. Every good ethnographer has had at least one episode where he or she felt awkward, uncomfortable, or threatened in the field, just as every bench scientist has at least on one occasion fumbled the setting on an instrument or mishandled a specimen. Rather than being deviations from acceptable practice, such missteps are *the essence of the experience*, giving the ethnographer insight into fundamental cultural differences, or the bench scientist a better understanding of her laboratory skills. Bench science, as many (Latour and Woolgar 1979, for example) have demonstrated, is a craft skill that one learns by doing, not simply by reading about it in a textbook. Craft skill, in contrast to industrial routine, engages the whole person in work that ultimately belongs to the craftsman, not the corporation (Sennett 2008). By exploring and extending the boundaries of craft skill and interpersonal engagement, one matures as a bench scientist or as an ethnographer.

Such personal engagement is necessary to demystify technology. When one has re-set a CONFIG.SYS file, or constructed a Web site, or rebuilt a clutch, or designed a piece of furniture, or fine-tuned a spectraphotometer, one has brought the object into one's own personal scale, just as a vernacular conversation with a villager recalibrates the cultural distance between the researcher and the villager. The distances and complexity remain, just as advanced hybrid engines are quite complex and some tribal rituals are quite confusing, but the distance and the complexity are no longer incommensurable.

The development cycle of a technology creates a set of social boundaries that are only understood through such exploration, the "process culture" that Constance Perin (2005) discusses. An experienced software developer will have an intuitive feel for when his beta version is ready for commercialization, based on experience with releasing not-yet-ready versions. By contrast, immature developers and immature software companies notoriously rush bug-ridden, unstable applications directly to market, without adequate field testing. Frustrated users pay for this every time their computers crash.

Cultural boundaries have this same character, in that they are located only through experience. One can locate the political boundaries of Japan on a map, but the boundaries of Japanese culture are understood only through engagement. An experienced ethnographer will have an intuitive feel for when she is about to cross a cultural boundary and tread lightly, in contrast to the tourists and adventurers who blunder around the world. Years of experience give the ethnographer a sixth sense of when she is approaching a cultural boundary, and heighten the antennae of sensitivity to cultural nuance. In both cases, bodily engagement, personal experience, and self-awareness are hard-won and critical parts of their knowledge, whether exploring the boundaries of technology or of culture.

In sum, the distance between the "Two Cultures" is not a difference between one intellectual topic or another. It is rather a difference between those who are *personally engaged* with living, breathing, physically embodied subject matter, whether it be tropical villagers, laboratory specimens, mechanical devices, or computer systems, and those whose knowledge of these matters comes only from reading about them.[1] Snow designated members of this latter group as the "literary intellectuals," yet today they include most of those whose only scientific contribution is science and technology policy.

Perhaps the greatest, unbooked externality of technology is the differentiation, complexity, and power asymmetries that it introduces into society, whether they are specializations in roles and industries, specializations in academic disciplines, or differences in wealth and style of life. This complexity, from a distance, can be as intimidating as the complexity of the instrument panel of a jetliner, as intimidating as the source code of a word processor, as intimidating as a text of criticism by Gadamer, or as intimidating as earthquakes and thunderstorms were in an earlier time. Feelings of inadequacy or helplessness are the most common reaction to such complexity and intimidation.

As social distances become greater, and social differences more extreme, a society loses the capacity for comprehension and compromise necessary for just and efficient government. Anthropology, with its unique capability for understanding and interpreting diverse subjectivities, should be able to lend some insight here. Whether technology diminishes or enhances our capacity for self-government is a question to which we will now turn.

Note

[1] Even source code, perhaps the most insubstantial of all technological objects, can be an object of intense bodily engagement, as any bleary-eyed, over-caffeinated developer will assure us.

Chapter Seven

Cultures of Technology

The revolution will not be televised. Observations and reflections may be submitted at http://www.allenbatteau.net

Contests between democratic and autocratic potential are found within any class of technology. Steam power was used to enslave men and women in dark satanic mills, and it was used to produce abundance for many. The Internet has been both a tool of global liberation and a tool of global surveillance. Technological standards create zones of autonomy and authority in which technological choices acquire the coercive force of Natural Law. The advance of technology, on the other hand, demands freedom to test the boundaries of accepted practice. Technologies as I have demonstrated translate not only matter and energy; increasingly they translate information and culture.[1]

In this chapter, I examine the technological translation of culture and the cultural translation of technology. I begin with the "information revolution" of the 1990s, which like earlier information revolutions—printing with moveable type in the 15th century, mass-circulation weeklies in the 18th century, photography and telegraphy in the 19th century, and electronic communication in the 20th—promised an advance of humanity into a sunny upland of universal enlightenment. I consider the various technologies that make up the most recent of these information revolutions, pointing out how they are mass-produced and have, themselves, become items of mass consumption. The role of these technologies in making culture into a commodity is examined. By so examining, I hope to show how culture is used as a vehicle for identity, thus becoming the most marketable of that which has been thought and said.

THE STRUCTURE OF TECHNOLOGICAL EVOLUTIONS

When the National Center for Supercomputing Applications at the University of Illinois in 1994 released Mosaic, the first graphical Web browser that evolved into Netscape, there was anticipation that this application would make the Internet more interesting to larger audiences. Prior to this time, the Internet, promoted by Vice President Al Gore as the "Information Superhighway,"[2] was a rarefied communication space for researchers. The World Wide Web, the protocols for which were first implemented by Tim Berners-Lee in 1990, was purely a text-based medium, in which highlighted text provided hyperlinks to other text-based Web sites with limited or no graphics (Berners-Lee, with Fischetti, 1999; Turkle 1995). The graphical interface had been created by Xerox PARC in 1973, adopted by Macintosh in 1984, and copied by Microsoft in the Windows operating environment, which, after several tries, achieved a reasonably stable version. Subsequently, millions of personal computer users, finding pictures more appealing than text, purchased Internet-ready computers, and began surfing the Web both for intellectual enlightenment and for pornographic excitement. As the World Wide Web turned into the World Wide Wait, with graphics-intensive files taking unbearably long minutes to download over 9,600-baud (*b*inary *au*dio *d*igit) modems and dial-up connections, telephone and cable TV companies raced to get high-speed connections to as many homes as profitable.

As university and military researchers found their private domain invaded by millions of first-time users on AOL ("the Internet with training wheels" in the view of experienced users), they began making plans for a even faster, more reliable, more exclusive "Internet2" that would be off-limits to the users of the "public Internet." As companies such as Amazon, Yahoo, and Google demonstrated that new business concepts could be used to generate excitement and that even the most questionable business models could attract venture capital (giving, for disappointed investors, fresh meaning to the 1993 *New Yorker* cartoon, "On the Internet, nobody knows you're a dog"). The "Internet Bubble" creatively destroyed billions in capital by 2000. Was this an information revolution?

In *The Structure of Scientific Revolutions*, Thomas Kuhn draws a distinction between "normal science," which proceeds incrementally, and "paradigmatic science," which proceeds discontinuously, creating through leaps of intuition new paradigms around which normal science then accumulates facts. For more than a thousand years, Ptolemy's geocentric universe was the reigning paradigm of Western astronomy. As observations improved and discrepant facts accumulated, the Ptolemaic paradigm became unwieldy until, in 1543, Nicolai Copernicus presented an alternative paradigm, which held that the earth and the planets orbited around the sun. This heliocentric paradigm, drawing on Arabic and Indian cosmological ideas, became the foundation of modern astronomy (Kuhn 1962).

A first reading of technological history suggests a similar pattern: the "air age," for example, began in 1903, and a century of refinements turned the Wright Flyer into a global utility. Marconi's demonstration of "wireless telegraphy" in 1896 launched a century of broadcast communication. However, this first reading is misleading: just as the Wrights incrementally improved on the experiments of Otto Lilienthal, Octave Chanute, and others, so too Marconi followed the efforts of others such as Pavel Schilling and Hans Christian Oersted, who, at the time, had been experimenting with electromagnetic radiation.

Technologies evolve, not so much by "improving" on earlier models, as by reassembling preexisting components, problems, and social groups. In a modern economy, new business models are part of the complexes driving technological evolution. Within any class of devices, there are numerous failed experiments as technologists experiment with alternative combinations of materials, architectures, and performance. Often what are called "revolutions" are more accurately labeled "performance breakthroughs" in which new levels of performance (speed, for example, or memory capacity or safety or pricing) lead to accelerated diffusion. In civil aviation, the performance breakthrough that turned commercial flight from a luxury good into an entitlement for the masses was the deregulation of the airlines in 1979, more than 50 years after Lindbergh crossed the Atlantic.

The error in looking for technological revolutions lies in assuming that a technology is contained in the device (a computer, a software application, a cell phone), without seeing the webs of social meaning in which devices are suspended. This is the view of the social construction of technology (SCOT) (Bijker et al. 1989), which examines technologies as networks of devices, social groups, social problems, and solutions, *each of which is capable of agency.* Each of these—devices, groups, problems—is capable of acting, of changing its state in response to environmental inputs and internal schema, sometimes in unpredictable ways. Technologies evolve, according to this perspective, by elaborating new configurations, exploring new spaces, and acquiring new meaning.

The bicycle, to use a famous example, evolved from the high-wheeled "ordinary," which was so difficult to control that it was mainly the device of daring young men (not acceptable for proper young ladies), into the "safety bicycle," with a diamond frame and two wheels of similar diameter. Along the way, there were experimental forays into numerous configurations, including the bicyclette, the Xtraordinary, and the geared Facile, and new drive mechanisms, including reciprocat-

ing levers and chain drives in addition to the direct crank of the ordinary. In considering how to create a bicycle useful to a wide range of
people and situations, new problems arose, including how ladies might
modestly ride a high-wheeled ordinary, how a bicycle could be controlled
on a rough road, or how a comfortable ride might be assured. New
groups beyond the daring young men became involved, including ladies,
the elderly, tourists, manufacturers, municipal authorities, and others.

Two features of this evolution should be noted: first, in the early
days, there was considerable *interpretive flexibility*—nobody knew what
or whom a bicycle was really for; the profusion of experimental configurations makes this clear. Like many technological devices in their
infancy, the bicycle started out as a mechanical device seeking usefulness. Second, as devices mature, a process of stabilization or *interpretive closure* sets in. The "bicycle" technology settled on the configuration
and uses that we are familiar with today, including racing, personal
transport, and delivery. Each of these, in turn, is associated with a distinct user community, including athletes, fitness buffs, and messengers.

Similar processes can be observed in every technology: for the
first decades of its existence, the airplane was seen purely as a weapons
and observation platform for the military, or a stunt platform for circus
performers. Only in the 1930s did commercial aviation become a viable
reality, not requiring direct subsidies. Similarly, the automobile
coevolved from a luxury good to a masculine device to an identity object
to a universal appliance, in concert with the evolution of production
techniques, women's roles, highways and service infrastructure, suburban housing patterns, and a mass-consumption economy. The current
configuration of suburban sprawl, which stabilized in an environment
of cheap energy, is now is a major inefficiency.

We might consider the evolution of technologies to be analogous to
the evolution of species, in which experimental subspecies either find
viable ecological niches or else die out, and in which the process proceeds not smoothly, but by fits and starts—punctuated equilibrium, in
Stephen Jay Gould's (2002) phrase. What appears *in media res* to be a
technological revolution is more accurately a moment of interpretive
flexibility, when new possibilities open up as fast as designers and
users can imagine them. The one difference in technological evolution
is that the niches for new technological alternatives are culturally
defined, in the same way that gender differences and gender-appropri-
ate behaviors are clearly a matter of cultural convention.

Two contrasting theoretical perspectives on this evolution are
technological determinism on the one hand, and the *social construction
of technology* on the other. Technological determinism views technologies as the driving force in history, evolving outside society according to
their own inner logic. As technology advances, in this view, society
adapts and evolves. Although this view accords with assumptions that

privilege technology and technological institutions, it has largely been abandoned by serious scholars of science and technology, being reserved for techno-enthusiasts of varying political stripes. The moments of interpretive flexibility in technological evolution have the implication that the trajectory of any given technology is more contingent, depending on numerous surrounding circumstances.

The social construction perspective, by contrast, focuses on the social antecedents and determinants of new devices. The example of the evolution of the safety bicycle given above describes the values, problems, and social groups that steered the safety bicycle toward its present configuration.

The social construction perspective, however, falls short in two respects. First, it confuses technology with tools and clever devices. In its infancy, the bicycle was *not* a technology within the understanding of technology that we have established. It was a clever device, and at times a useful tool, and considerable ingenuity went into the invention of its various configurations. But until standards were established, such as those for pneumatic tires in 1888, there is no advantage to considering it a technology rather than a clever device. Standards (such as those for pneumatic tires) *do* create zones of autonomy, so that a maturing technology *can* come to be a determining force within society.

Second, social construction needs to be balanced with a perspective of coevolution, in which technologies and social forms mutually adapt. Technological evolution is a complex adaptive process in which, like all evolving complex adaptive systems, (Holland 1995; Kauffman 1995), new and unexpected patterns can emerge. A substantial body of science has shown that adaptive systems, whether protein networks or regional climates or flocks of birds, exhibit collective behavior in undetermined ways. These systems are made up of thousands of autonomous agents whose behavior is guided both by their internal schemata and by their response to other agents in the environment.

Unlike some of the classic examples of complex adaptive systems, sociotechnical networks have power differentials lurking within them, differentials that create asymmetries in adaptation. A society can privilege certain groups, meaning that they are *not* expected to adapt to a changing environment: university professors, for example. Likewise, gender asymmetries are privileged and given shape and sturdiness within technology. Equally so, technology can privilege certain cultural assumptions, such as the *laissez-faire*, "free" market assumptions of contemporary political discourse in a technological world that demands massive planning and state intervention. Another privileged cultural assumption, having far-reaching technological consequences, is the medical model of illness, in which specific ailments call for specific, easily packaged interventions. We might also observe that societies can privilege certain problems: Pinch and Bijker (1989) observe that the "speed problem" was one of the driving elements in the bicycle's evolution. Might we not observe that the "speed problem" is privileged in the modern world, where *faster!* has become a categorical imperative (Gleick 1999)? For many decades of automotive evolution, the "speed problem" and the "styling problem" were in the driver's seat, and the "safety problem" was an afterthought, while the "tailpipe emissions problem" was ignored altogether, until federal legislation mandated catalytic converters. When one single problem is privileged, it forces technological evolution to adjust.

Anthropologist David Hakken argued in 1999 that the "computer revolution" was vastly overinflated, inasmuch as it had not materially altered relationships of production or created fundamental changes in consciousness. As an alternative to the advertising-driven rhetoric of the "computer revolution," he reviewed the ethnographies that had thus far been written about the effects of advanced information technology on work environments and social relationships. He found that while the rapid growth in use of computers in the 1990s did alter some labor processes, it did not create a fundamentally new type of social formation. As Bryan Pfaffenberger has argued (1988a), early proponents of personal computing fell back on an existing stock of social meanings and strategies and hence were constrained in their ability to visualize this technology in radical terms. Some people concluded: "The Internet is all hype."

So what was created?

DISCONNECTIVITY

One feature of our wired New Age is that we are connected like never before: connected to friends and colleagues on all five continents (or six, if you wish to ping the Amundsen-Scott South Pole Station in Antarctica), all 24 time zones, and in every major language group; connected to overwhelming sources of information, misinformation, and disinformation; and connected to an abundance of commercial and political opportunities to buy products, sign petitions, express opinions, and shout at the world. Social networking sites introduce users to new "friends"; e-mail brings us dozens of messages every day including inheritances from long-lost cousins in Nigeria; text messaging permits short bursts of asynchronous messages, and Twitter improves on this by making them synchronous.

It all started with cell phones and music-sharing devices. Indeed, one strongly could argue that portability provides the keys to the kingdom and that music is the currency of this realm. As musical experiences became disconnected from performance and even from physicality (vinyl, 8-tracks, CDs, live concerts), music became a free-floating currency that anyone could download onto his or her computer, iPod, cell phone (in the form of a ringtone), or personal Web site. Crysta Metcalf, Christine Miller, and Elaine Huang (2006), anthropologists working at Motorola, found in a study of cell-phone users that the exchange of music and ringtones was a form of gift giving, which like other forms of reciprocity, cements social relationships (Mauss 1990). YouTube was a logical extension, which permitted the sharing of video files, with the result that anyone with a cell phone with the necessary features could become famous for 15 minutes.

"Connectivity" and "information" have become the touchstones of the wired age, with little reflection on the fact that in themselves they

Technologies of connectivity, including cell phones, text messaging, and e-mail, can disconnect users from others close at hand.

have no value. Information, as contrasted to knowledge, is simply a deviation from randomness, but without a context, it conveys no understanding of the world. Connectivity also requires a context: every meaningful relationship in our lives includes not only the immediate parties to the relationship but a larger circle with which we share the relationship. Unshared "relationships" are of no social consequence.

The contexts of connectivity can be rich or bare, engaging or alienating, liberating or enslaving. All of the different components of communication—words, tone of voice, body language, phrasing, synchronicity, and proxemics—are made optional with the new devices of connectivity: with YouTube, for example, two individuals can "share" an experience without the inconveniences of proximity; with e-mail I can "share" my thoughts with you without being bothered by your response. In short, if I wish, I can dispense with all of the context-creating components of communication—multiple channels, consociates, social presence, knowledge objects—and come close to achieving a context of no context.

The comments of John Seely Brown and Paul Duguid (2002) (quoted in chapter 5) on decontextualizations—breaking society "down into its fundamental constituents, primarily individuals and information"—perversely reach their fullest potential with these technologies of connectivity. As an alternative to old-fashioned, face-to-face personal engagement, many people now use e-mail and blogs for angry, fire-and-forget missives, voice mail to avoid engaging in conversation, and surfing familiar Web sites to keep informed about the "real world."

The technologies of personal communication, in other words, create new opportunities for "disconnectivity"—avoiding engagements among contemporaries and consociates. As we become more connected with colleagues in China, we become less connected with our neighbors down the street; as we gain an ability to understand competing cultures around the world, we lose the ability to understand (and compromise with, and find common purpose with, and build common projects with) competing ideologies within our own nation. Edward Blakely and Mary Gail Snyder, in *Fortress America* (1997), have observed that with the growth in residential segregation (of which gated communities, the touchstone for their title, are but one manifestation), American public life has become more fragmented. With increasing class divisions, residential segregation, and occupational specialization, Americans are less likely than ever before to encounter, let alone be on familiar terms with, anyone who differs from them in educational attainment, social class, or political persuasion. The polity has evolved into numerous micropublics—of striving middle managers, earnest academics, angry conservatives, neighborhood activists, government technocrats, and a thousand others, each with their own circles of acquaintances, their own standard opinions, their own Web sites, and their own styles of individuality. Perversely, the Internet may be a great enabler of this fragmentation.

Dis-connection and re-connection are the metabolic rhythm of a dynamic culture. The *reconnectivity* of the information age is found in the emergence of online or "virtual" communities, heralded by Howard Rheingold's *The Virtual Community* (1993) mentioned in chapter 5. Samuel Wilson and Leighton Peterson in "The Anthropology of Online Communities" (2002) correctly observe that denoting online collectivities as "communities" brings in an idealized view of "real" communities, as (supposedly) places of boundedness, coherence, and mutuality. Two more realistic views of community that might be applied to Internet communities are first, Benedict Anderson's (1983) concept of "imagined communities," and second, Etienne Wenger's (1998) "communities of practice." For Anderson, an "imagined community" can consist of thousands or even millions of individuals who have never met, yet they are united, first by some common characteristic such as nationality, and second by a unique medium of communication, such as a newspaper in their own language. Imagined communities such as Germany, or even France, gave rise to the nationalisms of the 19th and 20th centuries (Anderson 1983). We might also note that as online sociality creates new forms of social connection as exemplified by Barack Obama's successful campaign for President in 2008, it can be expected to create new, emergent political formations.

"Communities of practice" for Wenger is a description of learning. A "community of practice" is a community of practical activity, in which individuals are drawn together by a mutual endeavor; it is also a community of learning-by-doing, in which individuals can try out and improve their skills in a supportive environment. A pressing question of the digital age is how computer users first learn computer skills and concepts. Although classes and instructional manuals are important, an equally critical part of the learning is from fellow users—a community of practice. Or, to make the point more strongly, individuals acquire technological literacy by joining networks of the technologically literate, and within these they find the mentoring, the conceptual orientation, and the opportunities for mastery and play required to get beyond the status of "clueless newbie."

Does all of this add up to a cultural change? I want to understand culture not as the commonplace parlance of "the way we do things around here" or as a term so imprecise that it encompasses everything people do. Rather, I want to understand culture in the two rigorous senses developed in the previous chapter: first as a community's shared resource for making sense out of the world, and second as personal achievement, cultivation, a life's work. In both senses, (Appadurai [1996:12] to the contrary notwithstanding), culture *is* reified and bounded by those who use it to define an identity. Culture is closely related to identity, and in neither sense is culture something that changes overnight with the introduction of a clever new device. In a cul-

ture of technology, change is marked not when a new device does wonderful things but when the device becomes iconic—a marker of identity.
The airplane, as a clever device, did not acquire broad cultural signifi-
cation until the 1930s (Wohl 2005), when heroic aviators such as
Charles Lindbergh, Italo Balbo, and Amelia Earhart became cultural
icons and Hollywood built motion pictures around aviation themes and
images. Avant-garde artists and science-fiction writers anticipated
this, but a breakthrough in mass consciousness of the airplane as an
iconic device cannot be dated before Lindbergh's 1927 flight to Paris.
Likewise, the emergence of the personal computer as a technological
icon can be dated to January 1983, when *Time* magazine placed it on
its cover as "Man of the Year."

Does the Internet represent a cultural change? The first decade of
the 21st century may turn out to have been the cusp of such a change,
in the same manner as the 1640s were the cusp of nationalism or the
closing years of the 18th century were the cusp of the Industrial Revolution. On the one hand, the fact that leading public servants could
have only a passing familiarity with "the Internets" (George W. Bush),
occasionally consenting to "do a Google" (John McCain), and the fact
that the iconic figure of the "Internet revolution" was a businessman
(Bill Gates) typifying a tribe of social misfits ("nerds"), one would conclude that this technology was still breaking through to mass acceptance. *On the other hand*, the election of a breakthrough political figure
(Barack Obama), on the strength of legions of Internet-connected vol-
unteers, would suggest that in 2008 the revolution arrived.

The breakthrough came with the community-building technologies of Web 2.0. The term "Web 2.0" was coined in 2002 to distinguish
user-content applications and portals (such as YouTube, Digg, and
Yelp; social networking sites such as Facebook and MySpace; and the
legions of nontechies creating blogs [Web logs]) from earlier applications (such as catalog shopping and making reservations online), in
which the Web was essentially an emulation of the old-fashioned
"dumb terminals" of hardwired networks. Web 2.0, by contrast, is constructed of content supplied by users, of user-assembled social networking, creating communities of common interest similar to the
manner in which the print-media business (newspapers and magazines) created the "imagined communities" of nationhood in an earlier
era (Anderson 1991).

The currency of the term "social networking" is revealing. As noted
in chapter 2, humans have been connected with each other in networks
for millennia. What is new in recent years is that the term "network"
has shifted from a commonplace assessment of the human condition
toward denoting asynchronous, computer-mediated many-to-many
communication platforms. The compound term "social networking" is
thus required to distinguish LinkedIn and MySpace from the (assuredly

social) networking of cocktail parties, religious revivals, and agricultural fairs that have existed for centuries. The "discovery" of networks (in exactly the same sense that Columbus "discovered" America) may be the beginning of a cultural New World (Castells 1996, for example).

Web 2.0, like every technology in its infancy—the airplane in 1908, the telegraph in 1845, the Newcomen engine in 1712—contains both democratic and autocratic potential, as the commons of a global village for enhanced civic awareness or as a new TV yet more appealing because it can be accessed anytime, anywhere, on cell phones and at office workstations. To the extent that users are in awe of its magical potential, the likelihood of those powers being enlisted by a knowledgeable elite are increased; to the extent that users are clear-eyed, understanding Web 2.0 not as a technology implying awesome mystery but as a space or an appliance for connectivity, its potential as a technology of freedom is enhanced.

THE COMMODIFICATION AND MEDIATION OF CULTURE

A culture takes shape in the media through which its collective representations are experienced. The Calvinist culture of 17th-century Geneva, with its abhorrence of adornment, had a different texture from that of the Catholic Church, with its profusion of icons, relics, robes, paintings, and statues. The culture of Sesame Street, with its fast-paced cutaways, its loosely coupled segments, and its cartoon-like puppet characters, has a different texture from that of the school classrooms found in institutionalized learning environments. In all cases, the questions one wishes to ask are what topologies of relationships are being created and what are the character and quality of these relationships: superficiality, intellectual engagement, sensuality, long-standing commitment, or short-term hookup? Is this a culture of liberation and commitment or a culture of manipulation and self-absorption?

Toward the close of the 20th century, American culture—in terms of frameworks for meaning and of personal cultivation—was dominated by television and its imitators, including the TV-inspired newspaper *USA Today*. In television, visual appeal, fast pace, discontinuity, and easily encapsulated thoughts ("sound bites") replace subtlety, reflection, and personal engagement; in a TV-dominated culture, an event is not "real"— that is, does not have moral or political consequences for the society— until its visuals have been seen on television. Ample demonstration of this comes from the differential impacts of the dozens of books that have been written on the use of torture and from that of a single photograph of a prisoner at Abu Ghraib. In this culture, not seen is not existing.

The communications researcher Neil Postman, in *Amusing Our-selves to Death* (1985), has described television as dominated by enter-tainment values, with the values of a typographic culture—extended discussion, reasoned argument, appreciation of complexity—crowded out by the discontinuities of programming that is interrupted every 12 minutes for commercial messages. The culture of "Now . . . this," which dominates the news, trivializes tragedy, placing earthquakes and mass killings on par with selections of flavors in mouthwash. Seemingly con-necting millions of viewers with the events of the day (the nightly news), with the spiritual (through televangelists), or with the affairs of state (through political speeches), television actually *disconnects* each of these from any historical, religious, or constitutional context. "The fundamental assumption" of the world of television, in Postman's view, "is not coherence but discontinuity" (1985:110).

In observing the discontinuity that TV promotes, Postman over-looks the continuities contained in its commercials. If we turn a televi-sion program inside out, and see it not as programming interrupted by commercials but as commercials interrupted by programming, we begin to see how television constructs its own reality. The only durable relationships that TV creates are between viewers and the brands that are advertised on TV. In fact, branding is the area in which coherence is most assiduously pursued. Advertisers take great care, spending mil-lions of dollars to assure that their brand identities engage viewers in the highest good of a consumer society: brand loyalty. Brand identities have far greater coherence and durability than the thoughtlets and sound bites presented as commentary on the TV news. The predomi-nance of brands as cultural objects is a consequence of electronic com-munication devices: in the 19th century, certain products were branded, but the disciplines of marketing, and the pursuit of brand loyalty, would have to wait until the emergence of consumer markets in the 1920s. Brand loyalty would reach its fullest potential with the technology, TV, that was most efficient in propagating these images.

Continuing with our inside-out exercise, we can see that in addi-tion to selling mouthwash and automobiles, TV is selling culture—cul-ture contained in recognizable characters such as the troubled teenager, in standard narratives such as the relentless detective fight-ing criminals (or, more recently, terrorists), and in landscapes of mem-ory such as the Old West (in *Star Trek*, relocated to interstellar space). That few if any viewers have any firsthand experience with the Old West is immaterial; the frontier and the Old West is such a strong image in America that it can even elect presidents. It has even provided a beguiling metaphor for the nation's science policy, as we saw in chap-ter 3, and a narrative for young entrepreneurs striking out on their own in California. This sale of culture is a complex transaction in which viewers devote their attention—their eyeballs—to the program, and networks aggregate eyeballs into "market share," which they then sell, for money, to fast-food chains, automobile manufacturers, sellers of personal-care products, and pharmaceutical companies. Images and eyeballs thus join cars, hamburgers, hair dye, and sleeping pills as being valued as commodities—for the fact that they can be bought and sold.

Culture as a commodity is, of course, easily as old as Barnum and Bailey, if not stretching back into the Middle Ages with the traffic in relics. Different forms and discourses have always existed in cultural spaces, with Great Books cultures, folktale cultures, family ritual cultures, and church member cultures always present as alternatives to commodity cultures. What marked TV in the late 20th century was how efficient and profitable commodity culture became, its profitability enabling it to rapidly expand into all available niches 24/7 around the clock, 360/90 around the globe.

There was, of course, pushback, such as Postman's book just noted, or Jerry Mander's *Four Arguments for the Abolition of Television* (1978). The rise of cable TV in the 1980s provided good evidence of dissatisfaction with the dominant networks, although the replacement of three "broadcast" networks with 300 "narrowcast" networks turned out to be no more than a continuation of the trends of disconnectivity that I have already noted.

The television culture of the 1960s, 1970s, and 1980s paved the way for the explosion of interest in the Internet in the 1990s. Although a long train of critics had been expressing dissatisfaction with television ever since Newton Minnow proclaimed it a "vast wasteland" in 1961, none of these tirades seemed to gain any traction as millions of Americans nodded off into passivity every night in front of the "boob tube." Yet, television, along with other graphic media, created a strong culture of visuality that throughout the 20th century replaced both the culture of orality and the culture of literacy. When the ability to access and communicate images from around the world at one's desktop first appeared in the mid-1990s, the acceptance was instantaneous.

In the vanguard of this acceptance, like that of prerecorded videocassettes a decade earlier, was the use of the Web to transmit pornography. The unregulated expanses of cyberspace, and the fact that it could be accessed in the privacy of one's own home (or hotel room), created the dynamic for a rapid growth in the transmission of pornography and similar sex-for-sale services, including prostitution. Whether this amounts to new opportunities for the degradation of women, or new channels for self-expression, depends in part on one's point of view. The Internet did not create the commercialization of sex, yet by creating new spaces beyond the reach of censors and state authorities, it established a robust business and technological platform that the industry previously lacked.

Perhaps nothing exemplifies the pushback against television as the meteoric success of YouTube. Starting out as a video sharing site in the spirit of Napster, it quickly became the way to capture the benefits of television including immediacy and contextuality but without the interruption of commercials, and with the advantage of selectivity and searchability. When it publicized a blatantly racist remark by a Republican Senator in 2006 (George Allen, calling a dark-skinned student a monkey), thus ending a rising star's political career, its status as a political force was confirmed.

Cultures and media always exist in competition and comparison with other cultures and media, even within the same community. For the Internet, many stodgy media companies wanted to treat it as another delivery vehicle for television. Early in the history of the Internet, and even today, a controversy exists over its topology: will it be a true network of many-to-many connectivity or a broadcast medium of one-to-many (see figure 2.1 on page 29)—in other words, a telephone exchange or a new sort of cable network for television? Cable TV companies, wedded to business models from the 1950s, favored the latter, and wired many households with Internet connections having greater capacity for downloading massive content than for uploading. Cable companies have requested permission to give certain commercial services greater bandwidth than others are given, thus making the Internet less of a democratic commons and more of a broadcast medium. Legislation currently in the Congress, upholding "net neutrality" for the time being, keeps the Internet equally open to all comers.

The driving force for the Internet in the 1990s was the thousands of entrepreneurs, some with questionable business models, some with no business models whatsoever, that rushed in to strike it rich, a phenomenon much resembling another California Gold Rush of 150 years earlier. For every breakthrough idea in a Google or an eBay, there were thousands of Internet sites that offered nothing more than catalog shopping or streaming "content." Just as Internet-as-TV vs. Internet-as-connectivity played out in legislation and in laying cable, the contest also played out in the competition among Web sites for viewers' content surfing and their interactive engagement.

A new set of platforms and applications is emerging out of this mêlée, which *may* turn out to be the appropriate technology of the Internet. YouTube, Facebook, Wikipedia, and Digg are all platforms for user-supplied content that, in 2008, created a groundswell of support for a presidential candidate, Barack Obama, who four years earlier was a virtual unknown outside his state senate district in Illinois. Obama's ability to coalesce an appealing image, an inspiring story, and a self-assembled network of millions into a triumphal election victory may mark a technological revolution.

TECHNOLOGY 2.0

Technologies, as I have demonstrated, have identities, identities that are created not only by their electromechanical characteristics but also by their social relationships and their embodiment of cultural distinctions and values. Lewis Mumford's identification of technological epochs exemplifies this, where he distinguished between the paleotechnical implements of the early industrial age and the neotechnical implements of the 20th century (Mumford 1934).

Culture cannot be understood without examining the social relationships within which it is embedded, relationships of both sharing and exchange, and relationships of authority and control. The dominant role within the culture of television is that of the consumer, a persona that in the contemporary economy has assumed an iconic status: television, newspaper, and magazine accounts anxiously examine what consumers are doing, and the current economic downturn is attributed, among other things, to a decline in "consumer confidence." The consumers, we are told, are shirking their duty to go out and buy more. Governments, corporations, and politicians spend billions of dollars to assure that consumers, a role now indistinguishable from that of "citizen," adopt the correct products, viewpoints, and candidates. The grand illusion of a consumer-oriented society is that the consumers are free to choose—between different brands of cigarettes, or different automobile body styles. Some things that consumers are not free to choose include electric cars, all-solar houses, or well-designed, well-maintained urban neighborhoods. Nor are consumers free to choose not to consume. These choices are either sufficiently scarce or entail sufficient social cost as to not constitute a "choice" in any empirical sense of the word.

An effort to inject choices into the social economy has been the movement for "appropriate technology," usually associated with the British economist E. F. Schumacher. A protégé of John Maynard Keynes and advisor to the UK Coal Board, Schumacher was closely acquainted with the opportunities and costs of modern industry. In *Small is Beautiful*, Schumacher argued in 1973 that large, corporate-sponsored technologies, such as energy-intensive factories or large-scale manufacturing, were unworkable in the Third World and actually did damage in places where they were supposed to help alleviate poverty. In contrast to C. P. Snow's (1998) argument for a leading role of science and scientists in alleviating world hunger, Schumacher argued that people in the Third World, when supplied with the tools, could do the job themselves. Schumacher's book inspired a broad movement of appropriate technology as an alternative development strategy, a push-back to the globalization that had hitherto defined "technology."

Appropriate technologies include bicycle-driven water pumps for arid regions lacking reliable electric supply or hand-cranked radios that never need to have their batteries replaced. They include minimally featured cell phones that are more reliable than landline telephones in many Third World locations. An ingenious application of appropriate technology principles is Loband, which simplifies Internet content (stripping out pictures, animation, and sound, leaving only text) and was developed by the NGO Aptivate, to bring the Internet to communities lacking high-bandwidth connections.

The distinguishing feature of appropriate technologies is not so much their cost or their engineered content as is their minimal externalities: their minimization of high-overhead inputs and infrastructure such as electric power supply, broadband connections, or a trained labor force, and their minimization of damaging outputs such as pollution and unemployment. Something as ordinary as building materials made from local resources, thus minimizing transportation expenses, can do more for rural development than a cement factory costing hundreds of millions.

Appropriate technology initiatives construct, on top of the existing, built environment, alternative devices, appliances, toolkits, and processes, such as electricity-collectors feeding back into the grid or modifications of automobiles to make them more suitable for the environment they are found in. Whether this is a "reverse salient" (a term that we introduced in chapter 3) or the beginnings of a new configuration cannot, at this moment of interpretive flexibility, be determined. Whether modularity and adaptability will constitute a true alternative to a technological world that hitherto has been marked by overwhelming size and power will be determined less in laboratories and more in political arenas.

The new Web applications may be creating just such a political arena. In contrast to traditional Web sites whose content was supplied by monolithic corporations or institutions, the spirit of Web 2.0 is user-supplied content and links: on Digg, for example, users can post, comment on, and "digg" (i.e., point to) news clips and videos that the community views, evaluates, and either promotes or demotes. With Facebook, one can learn more about casual acquaintances as well as keep up with friends. On Yelp, information about retailers, restaurants, products, and other items of commerce are created and shared by a growing community. Taken together, these and similar applications apply and leverage the "wisdom of crowds" and have even made their way into the traditional world of electoral politics.

Any prognostication of the eventual results of these new platforms would be as imprecise and as fatuous as a prognostication of the eventual results of Marconi's "wireless telegraph" 120 years ago. Whether technological literacy has sufficient widespread diffusion to permit the emerging community to control the course of Web 2.0 or whether its control will pass out of the hands of nerds and hobbyists and into the corporate realm is the technological drama of the current moment.

In sum, the Internet, although not creating a sunny upland of universal enlightenment, has opened up new spaces for all of the contradictory forces of a capitalist economy: new spaces for marketing, new spaces for the immiseration of labor, new spaces for consumerism, as well as new spaces for networking, for protest, and for subversion of consumerism's dominant ethos. Decisions about using and controlling these new spaces are fought out not only in the halls of Congress and in corporate boardrooms but also within entrepreneurs' pipe dreams, hobbyists' basements, social misfits' experimenting, political movements' social networking, and desktop users' decisions to move from lurking to participating in Internet communities. Like the first stirrings of user empowerment on the Internet in the 1980s, the results will not be televised.

Notes

[1] The concept of "translation" is useful here. In physics, translation is the movement of an object through Cartesian space. In linguistics, translation is the movement of discourse from one speech community to another. In both cases, there is a recontextualization of the object in motion.

[2] The vice president gave a nod to his father Al Gore, Sr., who, as a senator from Tennessee in the 1950s, was a leading sponsor of the Interstate Highway System, an automotive-age unification of the American nation.

Technology *for* Culture

In the modern world, technology and culture do not lead separate lives. As I have demonstrated, they are intimately connected, and just as our culture is increasingly defined by technological objectives, our technology is bounded by cultural capabilities. Ignorance of culture limits technological projects to narcissistic displays, and likewise, ignorance of technology confines cultural discourse to marginalia.

In this concluding chapter, I wish to reflect on what we have learned and pull together diverse strands of value, of networks and their topologies, of evolution and diffusion, and of identity, into a unified, articulated framework with which we might interrogate technology and culture. Although many techno-enthusiasts find no reason to question technology, I have demonstrated that many technological developments over the past century have made the world a more dangerous place and have purchased security and prosperity in core regions with hazard and want on the periphery. To reestablish a rightful relationship between technology and humanity, in which technology is a servant and humanity is the master, will require an effort of cultural reclamation, beginning with imagining a posttechnological society.

THE POVERTY OF UTILITY

Every object contains multiple elements of value, arrangements of significance that guide its social circulation. A sonnet has one sort of value, a digging stick another, and a John Deere tractor yet another. Some of these objects, like digging sticks, are quite simple, whereas others, like Shakespeare's sonnets, cram considerable complexity into 140 syllables.

In the ideology of technology, utility or usefulness is supposedly a guiding value, although I have demonstrated that the universe of objects that the modern world calls "technological" contains far more values as well: usefulness much of the time, but also social connections, magic, inspiration, and sublime beauty, *not one of which can be reduced to utility*. If our only understanding of technology is utilitarian, then our resources for controlling it are impoverished, and we diminish our ability to keep technology in its rightful place as a servant, rather than a master.

Paradoxically, techno-enthusiasts do recognize magic and sublime beauty in technology, even as they insist that it is only about usefulness. In their discourse of rationality, magic is irrational, and beauty is only decorative. Yet magic, understood as a social appreciation of awesome transformations of matter and energy, is real, and beauty, as any lover knows, is engaging. Even if it is only in the eye of the beholder, beauty, including the beauty of a hot car or a sleek cell phone, cements relationships that connect people both to each other and to the objects of common identities. These relationships are not trivial, but rather they are constitutive of a larger social order. Understanding how technological artifacts bind people to each other and to the artifacts themselves, either in attraction or repulsion, requires a more subtle mental exercise than techno-enthusiasts' wide-eyed wonder. The beauty and magic of technological objects come from their great condensation of art and social meaning within a simple package.

Technological objects also empower their users with a sense of mastery, although the ability of servants to enfeeble their masters is a very old story. Masters who assume that their servants are ignorant immediately render themselves vulnerable to the servants' schemes and manipulations. Likewise, the users of technology are trapped when they fail to appreciate that their attachment to their technology goes far beyond utility or personal efficacy.

CULTURE AT WARP SCALE

Ever since Einstein established that no object could travel faster than the speed of light, science fiction writers have imagined technologies that would permit just that. The favorite of these technologies folds or "warps" the space–time continuum along a fifth, non-Cartesian dimension, so that the astronaut's starship jumps from one location in the universe to another. "Warp speed, Mr. Spock," is a command to change the shape of the universe.

Every culture has its own space–time continuum, its own coordinates of relationships and boundaries of exploration and exploitation, measured by its technological development and its local circumstances. In some of the valleys of the New Guinea highlands, small villages are confined to single valleys, and the villagers never venture outside these landmarks; whereas, in the Canadian arctic, nomadic hunters range over thousands of square miles of land and sea. In some communities, men and women maintain intimate relationships within only a narrow circle, whereas today teenagers can twitter thousands of "friends" from Facebook or MySpace. In many agricultural communities, social relationships are measured by the length of travel they entail: one circle of acquaintances lives within an hour, whereas another is a day away. In every case, these boundaries are as much social as physical: the tree line in northern Canada marks not only the boundary between tundra and forest but also the boundary between Inuit and Athabascan.

In every 20th century lifetime, the coordinates of relationships and boundaries of exploration, even if only in the imagination, were changed multiple times, whether with automobiles, radio, television, the cinema, air transport, space exploration, or the Internet. The 20th century demanded that people learn new ways of relating, new expectations of their contemporaries and consociates, and new ways of being a citizen. In what shape and form does culture survive when its familiar contours are folded and warped with every succeeding generation?

Matthew Arnold saw "anarchy" as the alternative to culture, and a long string of writers ever since have echoed his dire warning that the world is on the road to ruin. From Matthew Arnold's Victorian vantage point, and that of his followers, adaptation was not a hallmark of humanity, least of all not adaptation to a horrifying world of industry. *Culture and Anarchy* was Arnold's defense of the ramparts against the rising tide of the debased standards of commercialism. By contrast, Arnold's contemporary, Charles Darwin, found adaptation and evolution to be sources of resilience and creativity in all species.

Technology reforms our social boundaries more durably than it expands our physical limitations. Even without voyaging to the moon, modern men and women have enlarged their range from a few hundred miles (a typical railway journey) to a few thousand miles (a trans-Atlantic trip, whether by airplane or ocean liner) in the course of a lifetime. Earlier long-distance trips, whether by immigrants to the New World or by pioneers settling the Great Plains or by English aristocrats on the Grand Tour, were more typically once-in-a-lifetime events prior to assuming a more settled existence. By the end of the 20th century, for large numbers of Europeans and Americans, ranging over thousands of miles became an ordinary event.

Social boundaries are defined not only by the circulation of our bodies and acquaintances but also by the circulation of our artifacts. The fact that I can eat fresh fruit from Mexico, and deliver a lecture via the Web to Ukraine, write these paragraphs using a processor from Tai-

wan, or supervise a student in Mumbai, represents a warping and fold-
ing of my social topology, a consequence of globalization far more
complex than any physical journey I might undertake.

Perversely, this expansion of range has magnified social distances, both within modern societies and around the world. Around the
world, this is obvious. A century ago, gentlemen at the Explorer's Club
could have some awareness of the Andaman Islanders, even if only in
Radcliffe-Brown's reports; today, by contrast, the military-industrial
complex has no clue why occupied countries would mount insurgencies
against their occupiers. Industrialists fluent in three or four modern
languages do business on four continents and converse easily with
their peers in the boardroom, even as they have less insight into the
conditions of their outsourced factory workers than did the local mill
owners of two generations earlier. Cell phone chatterers can walk down
the street oblivious to the street life around them. Technological acceleration, as I established in chapter 3, creates peripheries that are more
durable by virtue of the investments made in the large-scale systems
that created them. The price of connectivity around the world is
enhanced disconnectivity closer to home and enhanced opportunities
for misunderstanding around the globe.

TECHNOLOGICAL TRIBES

Part of the durability of technology-based social divisions comes
as well from the social investment in them. When early automobile
enthusiasts more than a century ago defined their lives—primarily in
terms of recreation and costume—around cars, it was little more than
a matter of amusement for everyone else in a horse-drawn society. Fast-
forward a hundred years to the end of the 20th century, and the automobile occupies a central place in American society and identity: urban
planning, domestic architecture, major industrial sectors, and large
swaths of mass consumption are arrayed around the automobile. Many
Americans further define who and where they are, their identity and
their social networks, in terms of their transportation devices: motor-
cycles vs. automobiles, family sedans vs. sporty convertibles, or fuel-
efficient hybrids vs. personal trucks.

David Hess' concept of "technototemism," which he first published
in 1995 and which I have reviewed in chapter 6, is thus amplified and
exaggerated, signaling not just consumer preference but social distinctions, codes for conduct, and consociation—identity, in short. This is not
identity that one can change at will, like a suit of clothes, but identity
that codifies and embodies the most important dimensions of one's life
in society. That Hess termed this "totemic" is important: totemic iden-

tities, unlike others, are nonnegotiable; they reference a "natural" order that is "outside" of society, whether the birds of the air and the beasts of the field in some tribal societies or the juggernauts of the road in our own.

The difference between a negotiable identity and a nonnegotiable identity is the difference between peace discussions and fighting words: between that where compromise is possible and that where it is not. Around the time of the Enlightenment, when Frederich von Schiller urged *alle menschen werden Brüder*,[1] there was a thought that modernity was erasing such nonnegotiable distinctions. A more accurate formulation, however, comes from the historian Peter Fritzsche. Observing how German nationalism, itself a nonnegotiable formation, used technology to define itself, Fritzsche stated:

> The histories of modern nationalism and modern technology are inexorably intertwined. Far from diluting nationalist passions, once thought to be ancient and mean, industrial prosperity and rational purpose gave them shape and sturdiness. Aviation, perhaps better than any other field of technology, clarifies the links between national dreams and modernist visions. (1992:3)

The fact that even the smallest countries feel a need for an air force and a national airline, that energy-rich nations develop nuclear industries, and that biological and chemical technologies, in addition to airplanes, are terrorists' weapons of choice, reinforces Fritzsche's conclusion. Whenever modern technology reinforces ancient and mean prejudices, whether modern nationalism or modern racism or modern fanaticism, then the entire justification for the modernist project of technological development is called into question.

THE SHAPE OF THINGS TO COME

In chapter 2, I identified the decisive role of 19th-century and 20th-century system builders in promoting technological evolution, and noted that their topological choices between centralized and networked systems have far-reaching social consequences. In chapter 7, I added the distinction of accessibility: the degree to which a system encourages and permits democratic engagement, vs. passive observation. The 19th-century railways were an example of a technology that was not open to civic participation, and the stranglehold they achieved on rural communities earned for them a nation's outrage. In the 20th century, broadcast television presented a similar monopoly, although open-access, decentralized technologies such as YouTube have undermined that monopoly, to no one's regret. Facebook, YouTube, and other decentralized, Web 2.0 technologies may well be the harbingers of a new Internet topology.

If my conclusion is correct, that access and topology are determinants of a technology's social character, then these questions add a new dimension to the discipline of technology assessment: to what extent does this technology embrace or undermine civic values? In technology assessment, questions such as what resources a new system requires or what performance can be expected of it, should be complemented with questions of who controls the technology, and to what purpose? Public investments in technology, such as those of the National Science Foundation or the Department of Energy, could then be guided by an understanding of whether a new technology is not simply useful and efficient, but what civic objectives are supported by its usefulness and efficiency.

Lest this seem too academic, in the current decade some of the biggest questions are over energy technologies. Within energy technologies that present alternatives to fossil fuels—solar, wind, geothermal, biomass, conservation—there are alternatives between small- and large-scale systems, corporate vs. citizen access, modularity vs. tight coupling, and centralization vs. decentralization. On paper, the large-scale, tightly coupled corporate systems will probably look more efficient, just as nuclear power did 50 years ago. If federal policy goes down this path, it will simply be purchasing one more burden for the economy and for our future community life. If, on the other hand, it goes down the decentralized, user-empowered Internet path—which, after all, did start out as a government program—it has the opportunity to create a citizen-engaged, robust, and ultimately more effective economy of energy supply, distribution, and consumption. Small-scale, modular systems have the added the virtue of diffusing more rapidly than large-scale, centralized systems with large corporate overheads. If one's objective is truly bettering lives through improving energy supply and distribution, then small-scale, appropriate technologies win every time.

CULTURAL RECLAMATION IN A POSTTECHNOLOGICAL SOCIETY

Nothing, we are told by Spinoza, is freer than the human imagination. Yet, one of the great tragedies of the technological society, as described by Ellul, is that for the majority of men it stultifies their imagination, making it more difficult for them to imagine alternative ways of life because, after all, the state of technology is so rational. Weber's "iron cage" of instrumental rationality is the ultimate project and product of the technological society. Despite Ellul's and Weber's views on the repercussions of technological instruments, we have seen that these amazing devices also have the capability to liberate the imagination, allowing everyone to imagine feats that were once considered delusional: trips to distant continents, instantaneous communication, mass destruction, and voyages to the moon. Can we enlist technology to imagine a posttechnological society, one in which technology is once again the servant and not the master of humanity?

It is clear that technology embraces many social values. Which values any given technology will embrace—autocracy vs. democracy, efficiency vs. beauty, and reverence for life vs. fascination with death—is an open question, particularly in the early stages of a technology's development. As a technology matures, it is instructed by its mentors and begins to take on their values. Personal computing, years after emerging out of the San Francisco counterculture, continues to have a rebellious streak, as exemplified by legions of hackers, whereas nuclear power encodes the massive centralization and apocalyptic vision of its fathers in the defense laboratories. All of this follows from the Laws of Society, and not the Laws of Nature.

A century of fascination with technology has given us the tools and vocabulary to begin imagining alternatives to the massive, energy-hungry, state- and corporate-oriented technological monuments that dominate much of the technological discourse. We can begin with the

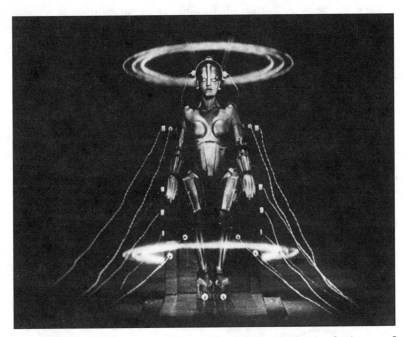

An early twentieth-century vision of the future. A robot in the image of a working girl in a modernistic city, from the black-and-white silent film, *Metropolis*, directed by Fritz Lang in 1926.

concept of appropriate technology, which hitherto has focused on Third World applications, but which may have value closer to home. Appropriate technologies are "appropriatable" technologies—devices and implements with which users can establish up-close and familiar relationships, so that mastering them no longer seems to be an insurmountable feat.

For every problem that technology is intended to address, other than the problem of space exploration, there is a nontechnological solution, one that focuses on remaking institutions and reforming the habits of the heart. For the "energy problem," conservation may well be the most efficient a solution, or more accurately part of an integrated solution. Conservation and efficiency standards would be more easily implemented than massive solar arrays, and more safely deployed than thousands of nuclear reactors. For the "immigration problem," legal reform and consensus building will require less federal budget and have more national benefit than a multibillion-dollar "virtual fence." For health problems, reform of institutions and daily habits will do more to add length and comfort to lives than any wonder drug on the horizon. The *problem* of technofixes such as virtual fences and wonder drugs, as I have demonstrated, is that they each come with hidden costs: massive complexity, massive surveillance, massive institutional inflexibility, and a potential for the atrophy of other capabilities.

This, however, is not an argument for abandoning technology, or for adopting "back to the earth" styles of life. Rather it is an argument for a clear-eyed, human-centered assessment of technology's opportunities, limitations, and hidden costs, and a mature integration of highly engineered implements and highly socialized objectives.

Note

[1] This expression, "All men become brothers," is from Schiller's *Ode to Joy*, which supplies the final chorus for Beethoven's 9th Symphony. "Joy, beautiful spark of gods / Daughter of Elysium, / We enter drunk with fire, / Heavenly one, your sanctuary! / Your magic binds again / What custom strictly divided. / All men become brothers, / Where your gentle wing rests." The *Ode to Joy* was, in 1989, made the anthem of the European Union, commemorating an end to hundreds of years of a continent divided by custom.

References

Adams, Henry
  1961[1918] *The Education of Henry Adams.* Boston: Houghton Mifflin.
Adams, Robert McCormick
  1996 *Paths of Fire: An Anthropologist's Inquiry into Western Technology.* Princeton: Princeton University Press.
Allison, Graham
  2004 *Nuclear Terrorism: The Ultimate Preventable Catastrophe.* New York: Henry Holt.
Anderson, Benedict
  1983 *Imagined Communities: Reflections on the Origin and Spread of Nationalism.* London: Verso.
Anderson, Benedict
  2001 *Imagined Communities.* New York: Verso.
  2002 *The Airplane, a History of Its Technology.* Reston, VA: American Institute of Aeronautics and Astronautics.
Appadurai, Arjun
  1996 *Modernity at Large: Cultural Dimensions of Globalization.* Minneapolis: University of Minnesota Press.
Apollonio, Umbro, ed.
  1973 *Futurist Manifestos.* Robert Brain, tr. New York: Viking Press.
Arnold, Matthew
  1994[1869] *Culture and Anarchy.* Samuel Lipman, ed. New Haven: Yale University Press.
Arthur, Brian
  1994 *Increasing Returns and Path Dependence in the Economy.* Ann Arbor: University of Michigan Press.
Asmann, Edwin N.
  1980 *The Telegraph and the Telephone: Their Development and Role in the Economic History of the United States: The First Century, 1844–1944.* Publisher not identified. University of Michigan Library.
Barley, Stephen R., and Gideon Kunda
  1992 "Design and Devotion: Surges of Rational and Normative Ideologies of Control in Managerial Discourse." *Administrative Science Quarterly* 37(3): 363–399.
Barnard, Chester
  1946 *The Functions of the Executive.* Cambridge: Harvard University Press.
Batteau, Allen W.
  1990 *The Invention of Appalachia.* Tucson: University of Arizona Press.
Bell, Daniel
  1947 "Adjusting Men to Machines. Social Scientists Explore the World of the Factory." *Commentary* 3 (January): 79–88.
  1999[1973] *The Coming of Post-Industrial Society.* New York: Basic Books.

134   References

Beniger, James R.
1986 *The Control Revolution*. Cambridge: Harvard University Press.

Bentham, Jeremy
2007 *Rationale of Judicial Evidence*. New York: Adamant Media. (Originally published, London: Hunt and Clarke, 1827.)

Berman, Marshall
1988 *All that Is Solid Melts into Air: The Experience of Modernity*. New York: Viking.

Berners-Lee, Tim, with Mark Fischetti
1999 *Weaving the Web: The Original Design and Ultimate Destiny of the World Wide Web by Its Inventor*. San Francisco: Harper.

Bigelow, Horace
1829 *Elements of Technology*. Boston: Hilliard, Gray, Little and Wilkins.

Bijker, Wiebe E.
1997 *Of Bicycles, Bakelites, and Bulbs: Toward a Theory of Sociotechnical Change*. Cambridge: MIT Press.

Bijker, Wiebe E., Thomas P. Hughes, and Trevor Pinch, eds.
1989 *The Social Construction of Technological Systems: New Directions in Sociology and History of Technology*. Cambridge: MIT Press.

Blakely, Edward J., and Mary Gail Snyder
1997 *Fortress America: Gated Communities in the United States*. Washington DC: Brookings Institution Press.

Boehm, Barry
1981 *Software Engineering Economics*. Englewood Cliffs, NJ: Prentice-Hall.

Braverman, H.
1975 *Labor and Monopoly Capital: The Degradation of Work in the Twentieth Century*. New York: Monthly Review Press.

Bray, T. L.
2000 Imperial Inca Iconography: The Art of Empire in the Andes. *RES: Anthropology and Aesthetics* 38: 168–178.

Brockman, John
1995 *The Third Culture: Beyond the Scientific Revolution*. New York: Simon & Schuster.

Brown, John Seely, and Paul Duguid
2002 *The Social Life of Information*. Boston: Harvard Business School Press.

Brynjolfsson, Erik, and Lorin M. Hitt.
2003 "Beyond Computation: Information Technology, Organizational Transformation and Business Performance," *Inventing the Organizations of the 21st Century*. Thomas W. Malone, Robert Laubacher, and Michael S. Scott Morton, eds. Cambridge: MIT Press.

Burke, Kenneth
1973 *The Philosophy of Literary Form*. Berkeley: University of California Press. (Originally published, "Semantic and Poetic Meaning." *Southern Review* 4 [Winter]: 501–523, 1938.)

Bury, J. B.
1932 *The Idea of Progress*. Mineola, NY: Dover Publications.

Bush, Vannevar
1945 *Science, The Endless Frontier*. Washington DC: U.S. Government Printing Office.

Carson, Rachel
1962 *Silent Spring*. Boston. Houghton Mifflin.

Castells, Manuel
1996 *The Rise of the Network Society*. Malden, MA: Blackwell Publishers.

Cornelius, David K. and Edwin St. Vincent
1964 *Cultures in Conflict: Perspectives on the Snow-Leavis Controversy*. Chicago. Scott, Foresman.

Cowan, Ruth Schwartz
1983 *More Work for Mother: The Ironies of Household Technology from the Open Hearth to the Microwave*. New York: Basic Books.

Cringely, R. X.
  1992 *Accidental Empires: How the Boys of Silicon Valley Make Their Millions, Battle Foreign Competition, and Still Can't Get a Date*. New York: HarperCollins.
Davenport, William H.
  1970 *The One Culture*. New York: Pergamon Press.
David, Paul
  2002 "Understanding Digital Technology's Evolution and the Path of Measured Productivity Growth: Present and Future in the Mirror of the Past." In *Understanding The Digital Economy: Data, Tools, and Research*. E. Brynjolfsson and B. Kahin, eds. Cambridge: MIT Press.
Deal, Terrence E., and Allan A. Kennedy
  1982 *Corporate Cultures: The Rites and Rituals of Corporate Life*. Reading, MA: Addison-Wesley.
Diamond, J.
  2005 *Collapse: How Societies Choose to Fail or Succeed*. New York: The Penguin Group.
Dizard, Wilson P.
  2008 "Senators Fume as FBI Admits Trilogy Foul-ups." *Government Computer News* (January 2).
Dobres, Marcia-Anne
  2001 Meaning in the Making: Agency and the Social Embodiment of Technology and Art. In *Anthropological Perspectives on Technology*. Michael Brian Schiffer, ed. Pp. 47–76. Albuquerque: University of New Mexico Press.
Downey, Gary Lee, Joseph Dumit, and Sarah Williams
  1995 "Cyborg Anthropology." *Cultural Anthropology* 10(2): 264–269.
Drucker, Peter
  1946 *The Concept of the Corporation*. New York: John Day.
Dutton, William, Brian Kahin, Ramon O'Callaghan, and Andrew W. Wyckoff
  2005 *Technological Innovation in Organizations and Their Ecosystems*. Cambridge: MIT Press.
Ehrlich, Paul
  1968 *The Population Bomb*. New York: Ballantine.
Ellul, Jacques
  1964 *The Technological Society*. Translated from *La Technique* by John Wilkinson (Paris, Librairie Armand Colin, 1954). New York: Random House.
Emerson, Ralph Waldo
  1994 "Ode, Inscribed to W. H. Channing." In *Emerson: Collected Poems & Translations*. Harold Bloom and Paul Kane, eds. New York: Library of America.
Emery, F., and E. Trist.
  1965 "The Causal Texture of Organizational Environments." *Human Relations* 18: 21–31.
Erikson, Kai T.
  1976 *Everything in Its Path: Destruction of Community in the Buffalo Creek Flood*. New York: Simon & Schuster.
Evans, Philip, and Thomas S. Wurster
  2000 *Blown to Bits: How the New Economics of Information Transforms Strategy*. Boston: Harvard Business School Press.
Feenberg, Andrew
  1999 *Questioning Technology*. London: Routledge.
Fritzsche, Peter
  1992 *A Nation of Fliers: German Aviation and the Popular Imagination*. Cambridge: Harvard University Press.
GAO (Government Accountability Office)
  2006 *Federal Bureau of Investigation: Weak Controls over Trilogy Project Led to Payment of Questionable Contractor Costs and Missing Assets*. GAO 06-306. Washington. Government Printing Office.
  2007 *Information Security: FBI Needs to Address Weaknesses in Critical Network*. GAO 07-368. Washington DC. Government Printing Office.

Gartner Group
1999 Travelling at the Speed of Hype—Gartner Group Predicts an End to E-business by 2008. (www.gartner.com/5_about/press_room/pr19991101a.html).
Gell, Alfred
1988 "Technology and Magic." *Anthropology Today* 4(2): 6–9.
Gibbon, Edward
n.d. *Decline and Fall of the Roman Empire*, Modern Library edition. News York: Random House.
Gibson, Kathleen
1991 "Tools, Language, and Intelligence: Evolutionary Implications." *Man* (NS) 26: 255–264.
Gillooly, Caryn
1998 "Enterprise Management Disillusionment." *Information Week* (February 16).
Gleick, James
1999 *Faster: The Acceleration of Just About Everything.* New York: Random House.
Goethe, Johann Wolfgang von
1963 *Goethe's Faust.* Walter Kauffman, trans. New York: Doubleday Anchor.
Gottinger, Hans-Werner
2003 *Economies of Network Industries.* New York: Routledge.
Gould, Stephen Jay
2002 *The Structure of Evolutionary Theory.* Cambridge: Harvard University Press.
Goyder, John
2005 *Technology and Society: A Canadian Perspective.* Peterborough, Ont.: Broadview Press.
Gras, Alain, Caroline Moricot, Sophie L. Poirot-Delpech, and Victor Scardigli.
1994 *Faced with Automation: The Pilot, the Controller, and the Engineer.* (Tr. from the French, *Face à l'automate.*) Paris: Publications de la Sorbonne.
Hakken, David
1999 *Cyborgs at Cyberspace.* New York: Routeledge.
Haraway, Donna
1997 *Modest_Witness@Second_Millennium. FemaleMan©_Meets_OncoMouse™: Feminism and Technoscience.* New York: Routledge.
Hardin, Garrett
1968 "The Tragedy of the Commons." *Science* 162(3859): 1243–1248.
Harvey, David
1990 *The Condition of Postmodernity.* Oxford: Blackwell.
Hauck, George F.
1983 "Hyatt-Regency Walkway Collapse: Design Alternates." *Journal of Structural Engineering* 109(5): 1226–1234.
Hendron, Alfred J., and Franklin D. Patton
1986 "A Geotechnical Analysis of the Behavior of the Vaiont Slide." *Civil Engineering Practice* (Fall): 65–130.
Hess, David
1995 *Science and Technology in a Multicultural World: The Cultural Politics of Facts and Artifacts.* New York: Columbia University Press.
Hobbes, Thomas
1668[1994] *Leviathan.* Edwin Curley, ed. Indianapolis: Hackett.
Hofstede, Geert
1980 *Culture's Consequences.* Beverly Hills: Sage.
1997 *Cultures and Organizations: Software of the Mind.* New York: McGraw-Hill.
Holland, John
1995 *Hidden Order: How Adaptation Builds Complexity.* Reading: Addison-Wesley.
Hughes, Thomas Park
1983 *Networks of Power: Electrification in Western Society, 1880–1930.* Baltimore: Johns Hopkins University Press.
1987 "The Evolution of Large Technological Systems." In *The Social Construction of Technological Systems.* Wiebe Bijker, Thomas P. Hughes, and Trevor Pinch, eds. Cambridge: MIT Press.

Hundal, Mahendra S., ed.
   2002 *Mechanical Life Cycle Handbook: Good Environmental Design and Manufacturing*. New York: Marcel Dekker.
Kaempffert, Waldemar
   1941 "War and Technology." *American Journal of Sociology* 46(4): 431–444.
Kauffman, Stuart
   1995 *At Home in the Universe: The Search for Laws of Self-Organization and Complexity*. New York: Oxford University Press.
Kern, Stephen
   1983 *The Culture of Space and Time: 1880–1918*. Cambridge: Harvard University Press.
Kern, Tony
   2001 *Controlling Pilot Error: Culture, Environment, and CRM*. New York: McGraw Hill
Kirby, L., ed.
   1990 *DNA Fingerprinting: An Introduction*. New York: Stockton Press.
Kling, Rob
   1999 "What is Social Informatics and Why Does it Matter?" *D-Lib Magazine* 5(1).
Knorr, Eric
   2005 "Anatomy of an IT Disaster: How the FBI Blew It." *Info World* (March 21).
Kuhn, Thomas
   1962 *The Structure of Scientific Revolutions*. Chicago: University of Chicago Press.
Kunda, Gideon
   1992 *Engineering Culture: Control and Commitment in a High-Tech Corporation*. Philadelphia: Temple University Press.
Langley, Samuel P.
   1891 *Experiments in Aerodynamics*. Washington, DC: Smithsonian Institution.
LaPorte, Todd, and Paula M. Consolini
   1991 "Working in Practice but Not in Theory: Theoretical Challenges of 'High Reliability Organizations.'" *Journal of Public Administration Research and Theory* 1: 19–47.
Latour, Bruno
   2005 *Reassembling the Social: An Introduction to Actor Network Theory*. New York: Oxford University Press.
Latour, Bruno, and Steve Woolgar
   1979 *Laboratory Life: The Construction of Scientific Facts*. Princeton: Princeton University Press.
Laurence, William L.
   1995 Nuclear History Due to Be Made in Geneva Parley. *New York Times* (August 6): 1.
Law, John
   2002 *Aircraft Stories: Decentering the Object in Technoscience*. Durham, NC. Duke University Press.
Leavis, F. R.
   1962 *Two Cultures? The Significance of C. P. Snow*. London: Chatto & Windus.
Lévi-Strauss, Claude
   1962 *La Pensée Sauvage*. Paris: Plon. (*The Savage Mind*. Rodney Needham, tr. Chicago: University of Chicago Press, 1966).
Lilienthal, David
   1944 *TVA: Democracy on the March*. New York: Harper and Brothers.
Lilienthal, Otto
   1893 *The Problem of Flight*. Washington, DC: Smithsonian Institution. (Annual report, 1894).
Mander, Jerry
   1978 *Four Arguments for the Elimination of Television*. New York: Morrow.
Marx, Karl, and Friedrich Engels
   1992[1847] *Communist Manifesto*. New York: Bantam Books.
Marx, Leo
   1964 *The Machine in the Garden: Technology and the Pastoral Ideal in America*. New York: Oxford University Press.

1997   "Technology: The Emergence of a Hazardous Concept." In *Technology and the Rest of Culture*. Arien Mack, ed. Columbus: The Ohio State University Press.

Mauss, Marcel
1990   *The Gift: The Form and Reason for Exchange in Archaic Societies*. W. D. Halls, tr. New York: W. W. Norton.

McCracken, Grant
2005   "When Cars Could Fly: Raymond Loewy, John Kenneth Galbraith, and the 1954 Buick." In *Culture and Consumption II*. Bloomington: Indiana University Press.

McCullough, D.
1968   *The Johnstown Flood: The Incredible Story Behind One of the most Devastating Natural Disasters America Has Ever Known*. New York: Touchstone Simon & Schuster.

McGroddy, James C., and Herbert S. Lin, eds.
2005   *A Review of the FBI's Trilogy Information Technology Modernization Program*. Washington, DC: National Academy of Sciences.

Means, James, ed.
1895–1897   *The Aeronautical Annual*. Boston: W. B. Clarke.

Metcalf, Crysta, Christine Miller, and Elaine Huang
2006   *Investigating the Sharing Practices of Family & Friends to Inform Communication Technology Innovations*. Report prepared for Motorola, Inc., Schaumburg, IL.

Miller, D., and H. Horst
2006   *The Cell Phone: An Anthropology of Communication*. Oxford: Berg.

Miller, Daniel
1987   *Material Culture and Mass Consumption*. London: Basil Blackwell.

Moore, Gordon
1965   "Cramming More Components onto Integrated Circuits." *Electronics* (April 19).

Mowery, D. and N. Rosenberg
1998   *Paths of Innovation: Technological Change in 20th Century America*. Cambridge: Cambridge University Press.

Mumford, Lewis,
1936   *Technics and Civilization*. New York: Harcourt, Brace.

National Commission on Terrorist Attacks.
2004   *The 9/11 Commission Report: Final Report of the National Commission on Terrorist Attacks Upon the United States*. New York: W. W. Norton.

Needham, Joseph
1954   *Science and Civilisation in China*. Cambridge: Cambridge University Press.

Nibourg, Theodorus
2005   "Conferencing Tools and the Productivity Paradox." *The International Review of Research in Open and Distance Learning* 6(1).

Noble, David F.
1997   *The Religion of Technology*. New York: A. A. Knopf.

Novak, Barbara
1980   *Nature and Culture: American Landscape Painting 1825–1875*. New York: Oxford University Press.

Nye, David E.
1994   *American Technological Sublime*. Cambridge. MIT Press.

Office of Technology Assessment (OTA)
1990   *Genetic Witness: Forensic Uses of DNA Tests*. Washington DC. Government Printing Office.

Ogburn, William F.
1922   *Social Change with Respect to Culture and Original Nature*. New York: B. W. Huebsch.
1941   Introduction. In *Technology and Society: The Influence of Machines in the United States*. McKee S. Rosen and Laura Rosen, eds. New York: Macmillan.

Ortner, Sherry
1973   "On Key Symbols." *American Anthropologist* 75(5): 1338–1346.

Perrow, C.
1984   *Normal Accidents: Living with High-Risk Technologies*. New York: Basic Books.

Perin, Constance
2005 *Shouldering Risks: The Culture of Control in the Nuclear Power Industry.* Princeton: Princeton University Press.
Petzinger, Thomas
1996 *Hard Landing: The Epic Contest for Power and Profits that Plunged the Airlines into Chaos.* New York: Three Rivers Press.
Peters, Thomas J.
1987 *Thriving on Chaos: Handbook for a Management Revolution.* New York. A. A. Knopf.
Peters, Thomas J., and Robert H. Waterman
1982 *In Search of Excellence: Lessons from America's Best-Run Companies.* New York: Harper and Row.
Pfaffenberger, Bryan
1988a "The Social Meaning of the Personal Computer: Or, Why the Personal Computer Revolution Was No Revolution." *Anthropological Quarterly* 61: 137–147
1988b "Fetishized Objects and Humanized Nature: Toward an Anthropology of Technology." *Man* 23(2): 236–252.
1992a "The Social Anthropology of Technology." *Annual Review of Anthropology* 21: 491–516.
1992b "Technological Dramas." *Science, Technology, and Human Values* 17(3): 282–312.
Pidgeon, N.
1997 "The Limits to Safety? Culture, Politics, Learning, and Man-Made Disasters." *Journal of Contingencies and Crisis Management* 5(1): 1–14.
Pinch, Trevor, and Wiebe E. Bijker
1989 "The Social Construction of Facts and Artifacts." In *The Social Construction of Technological Systems: New Directions in the Sociology and History of Technology.* Wiebe E. Bijker, Thomas P. Hughes, and Trevor Pinch, eds. Cambridge: MIT Press.
Postman, Neil
1985 *Amusing Ourselves to Death: Public Discourse in the Age of Show Business.* New York: Viking.
Powell, John Wesley
1889 The Lessons of Conemaugh. *The North American Review* 149: 150–156.
Psenka, Carolyn
2008 *A Monumental Task: Translating Knowledge in NASA's Human Space Flight Network.* PhD Dissertation, Wayne State University.
Rabinow, Paul
1996 *Essays on the Anthropology of Reason.* Princeton: Princeton University Press.
Rheingold, Howard
1993 *The Virtual Community: Homesteading on the Electronic Frontier.* Reading. Addison-Wesley.
Robbins-Gioia, LLC
2002 "ERP Survey Results Point to Need for Higher Implementation Success." *Business Wire* (January 28).
Roberts, Karlene
1990 "New Challenges in Organization Research: High Reliability Organizations." *Industrial Crisis Quarterly* 3: 111–125.
Roberts, Karlene, and Denise M. Rousseau
1989 "Research in Nearly Failure-Free, High-Reliability Organizations: Having the Bubble." *IEEE Transactions on Engineering Management* 36(2): 132–139.
Rogers, Everett.
1995 *Diffusion of Innovations.* 4th ed. New York. Free Press.
2003 *Diffusion of Innovation.* 5th ed. New York. Free Press.
Rogers, William B.
1861 *Objects and Plan of an Institute of Technology, Including a Society of Arts, a Museum of Arts, and a School of Industrial Science, Proposed to Be Established in Boston.* Prepared by Direction of the Committee of Associated Institutions of Science and Arts. Boston: John Wilson and Son.

Sahlins, Marshall
  1961 "The Segmentary Lineage: An Organization of Predatory Expansion." *American Anthropologist* 63(2): 322–345.
  1972 *Stone Age Economics.* Chicago. Aldine-Atherton.
  1976 *Culture and Practical Reason.* Chicago: University of Chicago Press.
Scarry, E.
  2003 *Who Defended the Country?* Boston: Beacon Press.
Schneider, David M.
  1968 *American Kinship: A Cultural Account.* Englewood Cliffs, NJ. Prentice-Hall.
Schumacher, E. F.
  1973 *Small Is Beautiful; ECONOMICs as if People Mattered.* New York: Harper and Row.
Schumpeter, Joseph
  1942 *Capitalism, Socialism, and Democracy.* New York: Harper and Brothers.
Scott, Judy
  2000 "The Fox Meyer Drugs Bankruptcy. Was It a Failure of ERP?" *Proceedings of the Association for Information Systems,* Fifth American Conference on Information Systems, Milwaukee, Wisconsin, August, 1999.
Sennett, Richard
  2008 *The Craftsman.* New Haven, CT: Yale University Press.
Shore, James
  2005 "It's Not Too Late to Learn." *SD Times* (August 15).
Shrivastava, P.
  1987 *Bhopal: Anatomy of a Crisis.* Cambridge. Ballinger.
Simpson, Matthew
  2006 *Rousseau's Theory of Freedom.* London: Continuum.
Snow, C. P.
  1998[1959] *The Two Cultures.* Cambridge: Cambridge University Press.
Stern, Gerald M.
  1976 *The Buffalo Creek Disaster.* New York: Random House.
Steward, Julian
  1955 *Theory of Cultural Change: The Methodology of Multilinear Evolution.* Urbana: University of Illinois Press.
Strauss, Lewis L.
  1954 Remarks Prepared for Delivery at the Founders' Day Dinner, National Association of Science Writers, Sept. 16, 1954.
Suchman, Lucy
  1987 *Plans and Situated Action: The Problems of Human-Machine Communication.* New York: Cambridge University Press.
Susman, Gerald I.
  1976 *Autonomy at Work: A Sociotechnical Analysis of Participative Management.* New York: Praeger.
Tarde, Gabriel
  1903 *The Laws of Imitation.* New York: Henry Holt.
Thompson, E. P.
  1963 *The Making of the English Working Class.* New York: Vintage.
Townsend, Peter
  1970 *Duel of Eagles.* New York: Simon & Schuster.
Trilling, Lionel
  1962 "Science, Literature, and Culture: A Comment on the Leavis-Snow Controversy." *Commentary* 33: 461–477. (Reprinted in Cornelius and St. Vincent 1964.)
Triplett, Jack E.
  1999 "The Solow Productivity Paradox: What Do Computers Do to Productivity?" *Canadian Journal of Economics / Revue Canadienne d'Economique* 32(2): 309–334.
Trist, E. L., and G. Susman, and G. Brown
  1977 "An Experiment in Autonomous Working in an American Underground Coal Mine." *Human Relations* 30: 201–236.
Turkle, Sherry
  1995 *Life on the Screen: Identity in the Age of the Internet.* New York. Simon & Schuster.

Turley, David
2000 *Slavery.* Oxford: Blackwell.
Vaughan, Diane
1996 *The* Challenger *Launch Decision: Risky Technology, Culture, and Deviance at NASA.* Chicago: University of Chicago Press.
Vidyarana, B. Gargeya, and Cydnee Brady
2005 "Success and failure factors in adopting SAP in ERP System Implementation." *Business Process Management Journal* 11(5): 501–516.
Vitruvius Pollio, M.
2004 *The Ten Books on Architecture.* Morris Hicky Morgan, tr. New York: Adamant Media. (Originally published, Cambridge, MA: Harvard University Press, 1914.)
Wajcman, Judy
1991 *Feminism Confronts Technology.* University Park: The Pennsylvania State University Press.
2000 "Reflections on Gender and Technology Studies: In What State Is the Art?" *Social Studies of Science* 30(3) (June): 447–464.
Wallerstein, I.
1974 *The Modern World System: Capitalist Agriculture and the Origins of the European World Economy in the Sixteenth Century.* Amsterdam: Elsevier/Academic Press.
Weick, Karl, and Karlene Roberts
1993 "Collective Mind and Organizational Reliability: The Case of Flight Operations off an Aircraft Carrier Deck." *Administrative Science Quarterly* 38: 357–381.
Weick, Karl, and Kathleen M. Sutcliffe
2007 *Managing the Unexpected: Resilient Performance in an Age of Uncertainty.* San Francisco: Jossey-Bass.
Weinberg, Alvin
1994 *The First Nuclear Era: The Life and Times of a Technological Fixer.* New York: AIP Press.
Wenger, Etienne
1998 *Communities of Practice: Learning, Meaning, and Identity.* New York: Cambridge University Press.
Westrum, Ron, and Anthony J. Adamski
1999 "Organizational Factors Associated with Safety and Mission Success in Aviation Environments." In *Handbook of Aviation Human Factors.* Daniel J. Garland, John A. Wise, and V. David Hopkin, eds. Mahwah, NJ: Lawrence Erlbaum.
White, Leslie
1949 *The Science of Culture, a Study of Man and Civilization.* New York: Grove Press.
Wiebe, Robert
1967 *The Search for Order, 1877–1920.* New York: Hill and Wang.
Wiesner, Jerome, and Herbert F. York
1964 "National Security and the Nuclear Test Ban." *Scientific American* 211(4): 27.
Wilson, Samuel M., and Leighton C. Peterson
2002 "The Anthropology of On-Line Communities." *Annual Review of Anthropology* 31, 449–467.
Wohl, Robert
2005 *The Spectacle of Flight: Aviation and the Western Imagination, 1920–1950.* New Haven, CT: Yale University Press.
Wolf, Eric R.
1982 *Europe and the People without History.* Berkeley: University of California Press.
Wynne, Brian
1988 "Unruly Technology: Practical Rules, Impractical Discourses, and Public Understanding." *Social Studies of Science* 18: 147–167.
Zachary, Pascal
1997 *Endless Frontier: Vannevar Bush, Engineer of the American Century.* New York: The Free Press.

Index